STEM by Design

Teaching with LEGO Mindstorms EV3

By Barbara Bratzel

Foreword by Chris Rogers

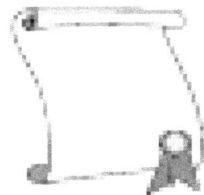

College House Enterprises, LLC
Knoxville, Tennessee

College House Enterprises, LLC.
5713 Glen Cove Drive
Knoxville, TN 37919, U. S. A.
Email   jwd@collegehousebooks.com
http://www.collegehousebooks.com

ISBN 978-1-935673-18-7

Foreword

As teachers, I believe our goal is to get students to move a story from our head into the student's head. The story could be what Shakespeare was saying in Macbeth, or it might be about the industrial revolution, or it might be understanding why things fall. We can do that through telling the students our story, lecturing them about gravity or Henry Ford, we can show them, through documentaries, laboratories, or demonstrations, and we can enable them to build their own stories, subtly nudging their growing understanding so that it agrees with current theory. The latter requires them to tell their own understanding, not repeat back ours. It requires them to argue, to find evidence, to understand alternate views. It requires them to "think like a scientist," weighing evidence, or "think like an engineer," balancing constraints. These are the same skills that they use when they are "thinking like an historian," reading accounts of both sides of a battle, or "thinking like a literary critic," finding evidence in the writing of others.

So the question then becomes how does one balance these three techniques? If we only lecture, students rarely get past a superficial learning, quick to repeat facts but with little understanding of what the facts really mean. If we only enable students to build their own understanding, then they are not taking advantage of the learnings of others over the last few centuries and will often make the same mistakes others made before them. The balance varies from student to student and a good teacher will be able to actively move her students from listening to watching to arguing and back to listening as they need it. Barbara is one such teacher, alternating between giving her students an overview of the material, showing them some examples, and letting them grapple with their own solutions in teams. In this book, she gives a number of examples in the STEM disciplines that will motivate the students to listen to the lectures and think about the demonstrations because it will impact their project.

One critical feature you will notice across all of the activities is that there is no right answer. Like she states in the book, a successful classroom is one where all teams have developed different solutions to the same problem - where they start to teach each other. She also points out that failure is an integral part of these classes, with kids being pushed to be creative, and therefore take risks, and therefore fail often on their way to coming up with a solution. In this way, I believe that one of the measures of the strength of a STEM class is by the range of solutions (solution diversity) at the end of class. A large range of successful products means that the problem was a good one, whereas if almost all the solutions look the same, then more than likely the students did not think independently.

This book is a great starting point for developing open-ended STEM classes, complete with solution diversity and creativity (and the failure that will come with it). Have fun being amazed at what your students develop; to me that is the best part of teaching.

Chris Rogers

Preface

A couple of months ago, one of my students recounted a conversation she had had with an admissions counselor at a school interview. She was telling the counselor about her science class, when the woman started asking her about specific robotics projects from the class—have you built music boxes yet? Cars without wheels? Robotic snails?

I was mystified. Who was this admissions counselor? An ex-student of mine? A past parent? How did she know about the class? It turns out that the counselor had no connection with my school. Instead, she had heard so much about the robotics class from students at interviews over the years that she was familiar with many of the projects we did.

That story made my day. I want to get kids fired up about STEM—and robotics and engineering design are powerful tools for doing that. Building something that actually works or writing a program that actually runs is exciting—and empowering.

STEM topics have been much in the news of late—the growing number of jobs in STEM fields, the dearth of women and people of color in STEM fields, the inclusion of engineering in the Next Generation Science Standards, the poor showings on tests of technological literacy among Americans young and old, the debate over whether every student should learn to code. Though people argue about the details, there is an emerging consensus that STEM education is vital—and lacking.

Robotics and engineering are great vehicles for teaching STEM concepts in an integrated way. They teach important skills as well: critical thinking, creative problem solving, collaboration, and communication. Working on projects shows kids that there is no single correct answer; that failure can be an important way of gaining knowledge; that perseverance, resilience, and flexibility, as well as technical knowledge, are vital to success.

This book is full of projects, large and small. Projects to teach programming. Projects to teach math and physics concepts. Projects to teach engineering design. Projects to teach kids to think creatively and work together to solve problems. My hope is that as the students do these projects, they will gain STEM knowledge, problem-solving skills—and enthusiasm.

Barbara Bratzel

Acknowledgements

Many people helped create the activities that make up this book. Most of all, I want to thank my students, whose insight, creativity, and humor have made my physics/robotics course such a joy to teach and whose feedback has strengthened the course considerably.

I am deeply indebted to Chris Rogers, without whom the course would not have been possible. He first introduced me to the idea of teaching science through engineering many years ago and he has been a constant source of inspiration, assistance, and infectious enthusiasm ever since.

I am grateful for all of the ideas and help I have gotten from the people at the Center for Engineering Education and Outreach at Tufts University, from the members of the science department at Shady Hill, and from the many dedicated and creative teachers with whom I have had the privilege of working.

Jim Dally, who edited and published all five of my books, improved my work immeasurably with his many thoughtful suggestions. Finally, this book would not have been possible without the patience and support of my husband, Jacob, and daughters, Audrey and Nell.

Table of Contents

Getting Started in the Classroom

About this Book

STEM by Design has seven sections. This first section, Getting Started, gives tips for managing the class and the materials. Part Two contains introductory activities for the EV3, both the hardware and the software. The third section contains a series of tutorials for teaching EV3 programming to a whole class with just one teacher. The structure of the tutorials lets the students work at their own pace and enables the teacher to concentrate on helping those students who need the most support. The fourth section is a series of activities for teaching STEM using the EV3. Part Five introduces the students to data logging. Part Six gives ideas for a number of creative engineering projects. The final section of the book contains low-tech labs, ones that use LEGO® pieces but not the EV3.

Materials

Software
The software used in this book is the EV3 Mindstorms software, an icon-based programming language. It is designed to run the EV3 programmable brick, but can also be used with the NXT brick, though not all of the features in the software are fully functional with the NXT.

LEGO Materials
The activities in this book are designed to be used with the EV3 Mindstorms Educational kit. This kit is similar to, but not identical to, the retail version.

Each pair of students in my class has an assigned EV3 Mindstorms set. In addition, the students can help themselves to extra LEGO materials from my collection. I have a bin for each type of LEGO piece—bricks, plates, wheels, gears and pulleys, etc.

My goal is never to have the availability of materials limit the students' ideas. After a number of years of teaching the course, I am near to this goal. Each year, I use my budget money to buy more of whichever pieces have been in short supply during the preceding year. However, some items, such as gears or wheels, can be a limiting factor. At the start of a project, I announce those limits to the class. For example, I might tell them that each group may only use two large wheels for the project.

I also have a labeled bin for each student group. The students store their projects in their bins between classes.

Useful additions: Most of the activities in this book make use of the parts found in the EV3 base kit. However, there are a number of pieces that are not found in the educational base kit, which are useful to have.

NXT Sound sensor: Not included in the base kit, but can be purchased separately.

NXT Temperature sensor: Not included in the base kit, but can be purchased separately. Especially useful for data logging.

Bricks, beams, and plates: Not included in the base kit, but useful for projects that involve a good deal of building, such as those in the Engineering Projects and Low-Tech Labs sections of this book.

Weight element (weighted brick): Useful for providing weight for ramp-climbing cars and other projects.

Computer Equipment

The ideal equipment for an EV3 class is one computer and one EV3 kit per group. Though the initial cost is substantial, the equipment can be used year after year. If one computer per group is not available, it is possible to do many of the projects with groups sharing computers. However, activities that require a substantial amount of programming are difficult with shared computers.

Another useful item is a projector or smart board, so that the entire class can view and discuss a program at once. This option is especially valuable for introducing new programming techniques and for viewing class data from experiments.

Classroom Management

Project Rules

I begin the year by distributing and discussing a set of project rules. The rules are included at the end of this section. I find that enforcing these behavior guidelines improves the tone in the classroom considerably.

Encouraging Help

From the first day of class, I encourage the students to collaborate and to help one another. No designs are secret—the students are welcome to incorporate good ideas from other groups into their projects. I stress that having another group adopt one's idea is a real compliment.

If someone finds a clever solution to a problem, I make that person the "expert" on that issue and send other students with similar problems to talk to him or her. If students are having a good deal of difficulty with a project and seem stuck, I encourage them to "go shopping," to wander around the class and look at everyone else's projects to see how others have tackled the problem.

I find that if I consistently give positive feedback to students who help one another, then soon the class culture becomes one of helping. The students no longer look to me as the only source of advice; instead, they consult one another. Aside from making the atmosphere of the class more pleasant, doing this means that I can supervise a whole class without being run ragged or having the students spend long periods of time waiting for help.

Competition

As part of my class, I run competitions all the time. However, I rarely have the students compete directly against one another. If a competition has a clear winner, it also has a room full of losers—not a cheerful prospect. Instead, I have the students compete against a set standard; for example, can you clear all of the LEGO® bricks from the field in fewer than thirty seconds. That way, the potential is there for the entire class to win. Also, with this kind of competition, there is no penalty for aiding others or sharing a good idea—helping another group succeed does not diminish your own group's prospects.

Failure

In a project-based class, failure is inevitable. An idea that seems good at first turns out to be a dead-end—or an outright disaster. As the upset students sit amid the ruins of their project, I praise them for taking a risk and point out the positive aspects of their plan. I help them see what they have learned about which designs do and don't work. I remind them that everyone in the class will have a design fail at some point in the year. I also make sure that they get no jeering or other negative response from the other students.

Appealing to Girls

Over the years, I have tried hard to make my course appealing to girls. The effort has been successful; girls and boys sign up for the course in roughly equal numbers. Of course, what appeals to a particular girl (or boy) varies widely, but I have found some general tendencies that have been useful in structuring activities.

One important lesson I have learned is that most girls do not like cars. Announce a project involving cars and the boys perk up while the girls' eyes glaze over. Because of this, I limit the number of projects I do involving cars, even though cars are a natural use of the EV3. Instead, I make fans, puppies, music boxes—projects that cover the same concepts but avoid cars.

Girls, in general, do not like direct competition as much as boys do. Instead of having the students compete against one another, I have them compete against a set standard. (See the section on competition, above.)

Girls tend to plan ahead more than boys. When I announce a new project, the girls grab a piece of paper and start sketching their ideas. The boys, on the other hand, head for the bins of LEGO materials to collect every piece they think they might need, then worry abut what they are going to do with them. To accommodate both styles, I have accumulated a lot of LEGO materials. When the girls finish planning and go to gather their materials, there are still plenty of choice LEGO pieces left.

I tend to pair girls with girls and boys with boys, especially at the beginning of the year. I have found that mixed-genders pairs sometimes struggle to work together smoothly, much more so than in traditional science courses. The biggest complaint: the girls say that the boys take over the building projects.

One last note: color matters. Girls tend not to like the stark white-and-gray color scheme of the EV3 as much as boys do. In fact, I find that the girls generally notice color more than the boys do. Girls are much more likely to plan a color scheme for their projects. I make sure that my LEGO® brick collection contains the full range of colors—purple, orange, pastels—as well as the traditional LEGO brick colors.

Builders vs. Programmers

Many of the students quickly find one aspect of the projects, either the building or the programming, more appealing than the other and start to specialize. I allow this to some extent, while stressing that everyone needs to become competent at both. To ensure that they do, I give periodic building quizzes and programming quizzes. Neither one is particularly difficult; instead, each is designed to test whether the students have mastered the basics.

Assessment

I use a wide variety of assessments in my classroom—rubrics, engineering notebooks, design sketches and explanations, student presentations and videos. (In addition to these project assessments, I give traditional problem sets and tests on the physics concepts.)

Whatever method I use, there are two pitfalls I try to avoid:

- Grading based on success—it discourages innovation, making the students more likely to "play it safe."
- Grading based on creativity—it is subjective and stressful (for both teacher and students).

Instead, I grade on documentation of the design, understanding of the concepts underlying the design, and systematic improvement to the design.

Project Rules

We will be doing projects, both large and small, throughout the year. Sometimes your attempt will be a success. Other times, it will be a miserable failure. Either result is okay—failure is an integral part of the design process.

What is not okay: laughing at someone else's project.

So, some rules for working in the class:
1. You may not criticize or make fun of anther student's work. This includes laughing, teasing, or comparisons ("My car is so much better than your car….").
2. However, you may provide CONSTRUCTIVE criticism. To be constructive, the comment must be specific and offer a possible solution. For example, "Hey your car doesn't go straight" is not acceptable, but "I noticed that your car veers to the left. It looks like the back wheel is rubbing against the frame" is welcomed.
3. No designs are private property. Anyone may get ideas from any other design. If someone copies a piece of your design, the proper reaction is to be flattered—clearly, the other person has recognized your brilliance.
4. If you get stuck, feel free to look at other people's designs to see how they have solved similar problems.
5. And finally, relax! Things will go wrong—but you will have plenty of time and assistance to fix the problems.

Part Two: Introductory Activities

This set of activities introduces the EV3 programmable brick and the EV3 Mindstorms software.

The activities in this section are:
1. Making Connections
2. Build a Box
3. Fancy Box
4. Pandora's Box
5. Sneak Attack
6. Baker's Dozen Car

Making Connections introduces the students to the peg-and-beam building system of the EV3 by having them investigate the differences among the various pegs. The next activity, Build a Box, gives the students practice using beams and pegs to build. In Fancy Box, the students improve upon the box by writing a simple program to display an image and play a few words. Pandora's Box and Sneak Attack introduce the use of a sensor to control an action. The last activity, Baker's Dozen Car, guides students in building and testing a simple car.

Making Connections

Your EV3 kit contains four different connector pegs, a long cream one, a long blue one, a short black one, and a short gray one.

The short pegs can be used to connect two beams; the long ones can be used to connect three.

Besides color and length, there is another difference among the pegs. Can you figure out what it is? Hint: try using the pegs to attach beams to one another.

1. Describe the third difference:

2. In terms of length, the cream and blue pegs are alike. How about in terms of the third difference you just found? Which colors are alike?

You also have red connector pegs, which can connect two beams and will also accept an axle.

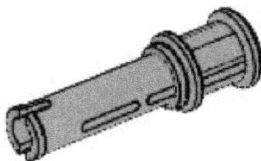

3. Which category do they fit into? How can you tell?

Teacher Information   Making Connections

This introductory activity allows the students to become familiar with the beam-and-peg building system and to learn the difference between friction and non-friction connector pegs.

Objectives
1. To gain comfort in building using beams and pegs.
2. To learn the difference between friction and non-friction connector pegs.

Materials
LEGO® beams and connector pegs

Time: Approximately 15 minutes

Notes
1. If the students have trouble spotting the difference between the pegs, have them use the pegs to attach beams to one another and then try turning the beams.

Answers to Making Connections

1. Some pegs allow beams to rotate freely. They are known as non-friction connector pegs. Others are a tighter fit; they do not allow beams to rotate freely. Those pegs, unsurprisingly, are known as friction connector pegs.
2. The long cream and short gray pegs are non-friction pegs that allow beams to rotate freely. The long black and short blue pegs are friction connector pegs that do not allow beams to rotate freely.
3. The red axle connector pegs are friction pegs, like the blue and black ones.

Build a Box

Build a box to hold two plastic balls. The box must use the EV3 as its base and must have a hinged lid for placing and removing the balls.

Here are some helpful hints:

Connector pegs can be used to attach a beam to the EV3. If a single peg is used, then the beam will be able to swing back and forth (a hinge!). If two or more pegs are used, the beam will be held in a fixed position.

Axle extenders can be used to attach two axles end to end to create a longer axle.

Extra: If you have time, design a latch for your lid so that it can be fastened shut.

1. What was the easiest aspect of building the box?

2. What was most difficult?

3. Make a sketch of your hinge. Label the LEGO pieces you used to build it.

Teacher Information   Build a Box

This introductory activity allows the students to practice using the beam-and-peg building system and to learn how to attach beams to the EV3.

Objectives
1. To gain comfort in building using beams and pegs.
2. To engage in creative problem solving.

Materials
EV3, LEGO® pieces, plastic balls

Time: Approximately 40 minutes

Notes
1. Plastic Ping-Pong balls or other similarly sized balls work well of this activity.
2. Stress that there are many possible solutions to this problem. At the end of the activity, have the students share some of their favorite ideas with the rest of the class.

Answers to Build a Box

1. Answers will vary.
2. Answers will vary. Common answers are designing the hinge or figuring out how to attach beams to the EV3.
3. Answers will vary. The sketches should be detailed enough that the reader could construct the hinge from the drawing.

Fancy Box

Open the EV3 Mindstorms software. From the File menu, choose New Project and then Program. After the new window opens, double-click on the Program tab and name your program.

You will see a white area with a Start block on it. This is where you will write your program. At the bottom of the screen is a green tab with a series of blocks. This is the Action Block palette.

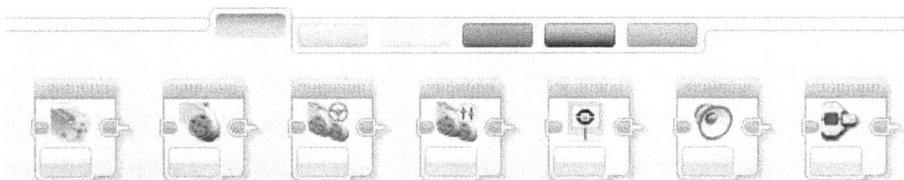

The fifth block in the palette is the Display block. The sixth is the Sound block. You will use both of these to make your box fancier.

Click on the Display block and drag it onto the grid next to the Start block. Drop the Display block. The two icons should connect together and the Display block should no longer be grayed out. Click on the folder on the Display block and choose Image. Next, go to the blank rectangle in the upper right corner of the block. The words File Name should pop up. Click on the box. A list of possible images will appear. Choose the picture that you want to appear on the floor of your box.

Now click on the Sound block. Just as you did for the Display block, drag it to the end of the program and choose a sound to play from the file.

Connect the EV3 to the computer using the USB cord. Turn on your EV3. Download your program to the EV3 and run it by pressing the Download and Run arrow in the lower right corner of the computer screen.

Notice that your image disappeared as soon as the sound was finished. This happened because the program ended as soon as the last block (the Sound block) executed.

Here's how to make your picture and sound last longer: Choose the Wait block from the Flow palette (the orange tab at the bottom of the program window), add it to the end of your program, and set the time for ten seconds.

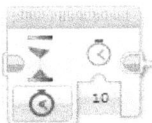

Next, go back to the Sound block and change the Play Type from 0 (wait for completion) to 2 (repeat). Download and run this new version of the program.

Teacher Information   Fancy Box

This activity introduces the EV3 software with a simple two-icon program.

Objectives
1. To learn how to open the EV3 software and write a simple program.
2. To download and run a program.

Materials
EV3 box from the previous activity, computer

Time: Approximately 20 minutes

Notes
1. You may want to demonstrate to the students how to start the EV3 software and open a new program before starting this activity.
2. You may want to show the students how to name and save a program.

Sample Program for Fancy Box (brief version)

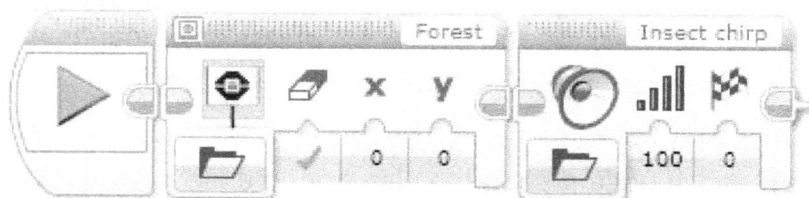

Sample Program for Fancy Box (extended version)

Pandora's Box

In the Greek myth, Pandora lets her curiosity get the better of her and opens a forbidden box, unleashing evils upon the world. To prevent unauthorized access to your box, attach a warning system to it, sounding an alarm whenever the lid is lifted.

To detect the lifting of the lid, you will use the gyro sensor, which can be used to detect tilt or angle. To do this, attach the gyro, resting on its side, to the lid of your box.

Write a program to make your EV3 sound an alarm whenever the gyro sensor detects a tilt.

You will need two blocks, the Wait block (located under Flow) and the Sound block (located under Action).

To set the Wait block to monitor the gyro sensor, click on the clock in the lower left corner of the block and select Gyro Sensor, then Compare, then Angle. Set the Compare Type to > or ≥ and choose what size angle will trigger your alarm. The number in the upper right tells you which port the sensor is connected to; plug your gyro sensor into that port.

To set the Sound block to sound an alarm, choose Play File in the lower left corner. In the upper right corner, click on File Name and choose LEGO Sound Files. Choose an appropriate word or sound for your alarm.

Turn on your EV3 and download your program, being careful that the gyro sensor stays absolutely still. (Otherwise, it may not calibrate correctly.) Test your program.

Extension: Change your program to make the alarm sound repeat several times rather than just playing once.

Teacher Information   Pandora's Box

This activity introduces the gyro sensor and the Wait block.

Objectives
1. To learn how to use the gyro sensor.
2. To use the Wait block.

Materials
EV3 box from the previous activity, gyro sensor, computer

Time: Approximately 20 minutes

Notes
1. Instead of using the compare function of the gyro sensor, one can also use the change function and wait for a relative change in position rather than an absolute movement.
2. It is important that the gyro sensor be absolutely motionless while it initializes. Holding it in one's hands, even if it is kept still, is enough to make it initialize incorrectly.

Sample Program for Pandora's Box

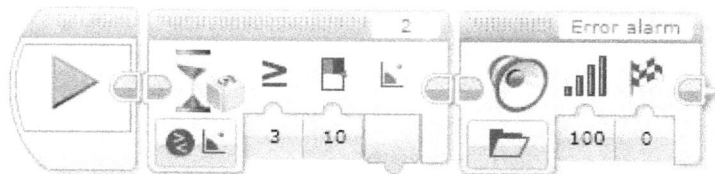

Sample Program for Pandora's Box Extension

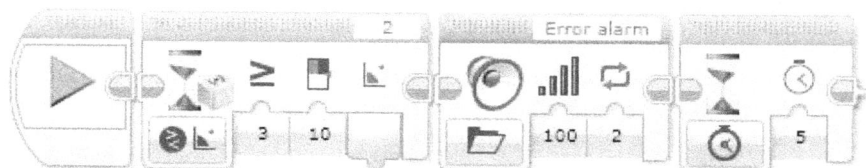

Sneak Attack

The gyro sensor has two different modes, angle and rate. In angle mode, the gyro sensor measures the actual angle in degrees. In rate mode, the sensor measures the rate of change of the angle.

In Pandora's Box, you used the gyro sensor in angle mode to make an alarm for the box. Now, you will use rate mode to create a game, Sneak Attack.

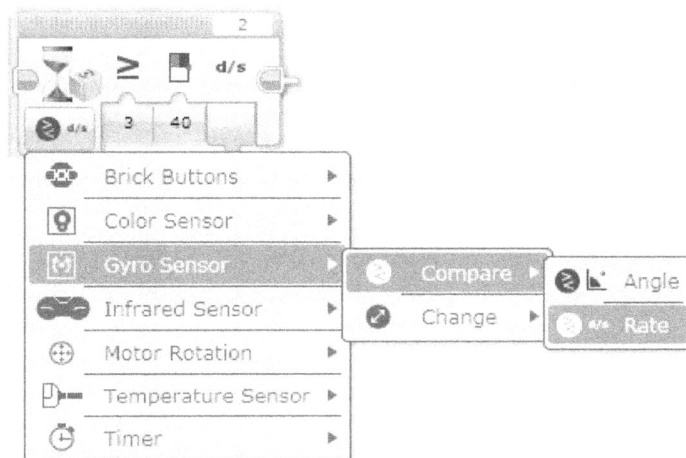

In Sneak Attack, the player has to move slowly enough to avoid setting off an alarm. The premise of the game is up to you—it can be removing a treasure from a box, opening a door to sneak into a room, prying open a crocodile's jaws without waking it, or anything you want.

Teacher Information   Sneak Attack

This activity introduces the rate mode of the gyro sensor.

Objectives
1. To learn how to use the gyro sensor in rate mode.
2. To gain additional building and programming practice.

Materials
EV3, gyro sensor, LEGO® pieces, computer

Time: Approximately 20 minutes

Notes
1. Just as in Pandora's Box, instead of using the compare function of the gyro sensor, one can also use the change function.
2. It is important that the gyro sensor be absolutely motionless while it initializes. Holding it in one's hands, even if it is kept still, is enough to make it initialize incorrectly.

Sample Program for Sneak Attack

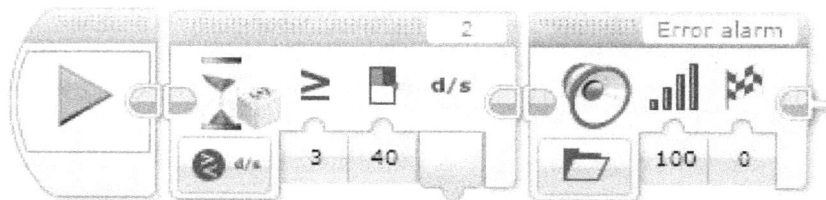

Baker's Dozen Car

Simplicity can be a wonderful thing.

Build a simple EV3 car using any or all of the pieces shown below—plus no more than thirteen additional pieces.

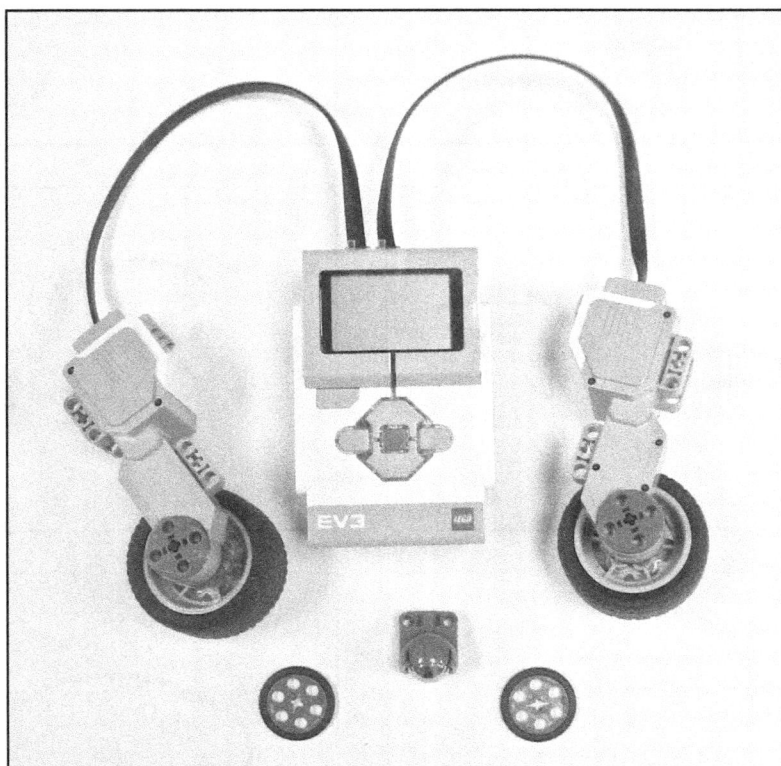

Your car must be able to go forward, go backward, spin right, and spin left—all without falling apart. There are many possible solutions to this challenge—there is no one "correct" answer. Feel free to tinker.

Test program:
In order to test your baker's dozen car, write an EV3 program that runs the car forward, backward, spin right, and spin left, each for two seconds.

To help get you started, here are some possible solutions. Each car shown uses no more than thirteen additional parts.

Teacher Information   Baker's Dozen Car

The students learn how to build simple EV3 cars.

Objectives
1. To build a two-motor EV3 car.
2. To gain additional experience with beam-and-peg construction.

Materials
EV3, motors, LEGO® parts including wheels, computer

Time: Approximately 40 minutes

Notes
1. You may want to highlight a few useful pieces for the students, such as the rectangular frames and the L-shaped and H-shaped beams with connectors.
2. A simple reasonably sturdy car can be built with as few as four additional pieces.

Sample Program for Baker's Dozen Car

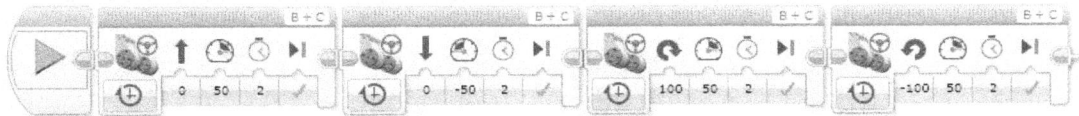

This program will run a car forward for two seconds, backward for two seconds, spin right for two seconds, spin left for two seconds. Depending upon the way the motors are oriented, the movements may be reversed: backward first, then forward, and so on.

Programming Sequences

Each of the following sets of activities is designed to be done as a single unit, with the students working at their own pace, receiving help from the teacher as needed. Each activity sheet contains a short program to write. Of course, the activities can be used as stand-alone activities as well.

One way of organizing the class as the students learn the EV3 Mindstorms programming language is to give them activity stamp sheets to fill out as they work. When a student successfully completes an activity, he or she brings the EV3 with the completed program to the teacher for inspection. The teacher approves it by stamping the student's activity sheet in the appropriate box. The student is then free to progress to the next activity. Using the activity stamp sheet allows the teacher to keep track of everyone's progress while concentrating on helping those students who need the most assistance.

Some groups will probably complete a programming sequence before others do. Each sequence can be followed by an open-ended activity. As students complete the programming sequence, they can start the follow-up activity. The Teacher's Information page for each sequence has suggestions for follow-up activities.

The programming sequences, each of which contain four activities, are:

- Seeing Red: Wait block, Sound block, Loop block, Switch block, measuring color using the color sensor.
- Driver's License: Move Steering block, Move Tank block, Start and Stop blocks, measuring light using the color sensor.
- Fan Club: Medium Motor block, Display block, Math block, data wires, loop index, nested loops, touch sensor.
- Traffic: Brick Status Lights block, Loop Interrupt block, Sound Editor, Image Editor.
- Number Sense: Random block, Text block, advanced uses of the Math block and loop index.

A last group of programming activities, Additional Programming Topics, addresses advanced topics. Unlike the previous tutorials, these activities are not designed to be completed in order. Instead, the teacher can pick and choose, using those activities that cover topics that will be most useful to the students.

These activities are:
- Cautious Car: Logic Operations
- Touch Tally: Variables
- Brick-button Navigator: Arrays
- Mail Delivery: My Blocks

Seeing Red

As you complete each programming activity, demonstrate your program to the teacher. Make sure that you get the stamp of approval for each activity before moving on to the next one.

## See and Say	
## Red, Red, Red	
## Red or Not	
## Rainbow Detector	

See and Say

The EV3 has several sensors, including a color sensor that can be used to detect different colors as well as light.

Write a program to make your EV3 see red. The robot should wait until it sees the color red, say "red," and end the program.

You will need two blocks, the Wait block (located under Flow) and the Sound block (located under Action).

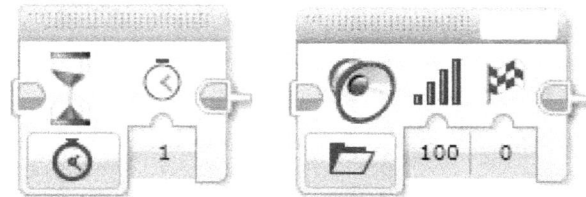

To set the Wait block to monitor the color sensor, click on the clock in the lower left corner of the block and select Color Sensor, then Compare, then Color. The number in the lower middle of the block tells you which color you will wait for (red is 5). The number in the upper right tells you which EV3 port the sensor is connected to.

To set the Sound block to say "red," choose Play File in the lower left corner. In the upper right corner, Click on File Name, choose LEGO Sound Files, Colors, and then Red. Download and test your program.

Red, Red, Red

Make your See and Say program repeat—after it sees and says "red," have the program go back to searching for more red. To do this, add a loop to your program. Any blocks that you place inside the loop will execute each time the loop runs.

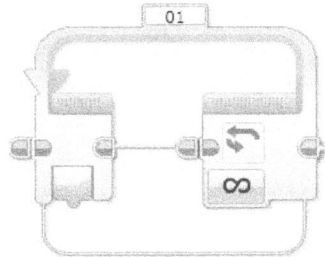

The loop icon shown above will run forever, as indicated by the infinity symbol, which looks like a sideways eight. You can adjust the loop to run for a certain number of times or to run until an event such as the push of a button occurs.

Adjust your loop to run until the bottom button on the EV3 is pushed. First, click on the infinity symbol and choose Brick Buttons. Then, select the bottom brick button (5) from the Set of Brick Button IDs.

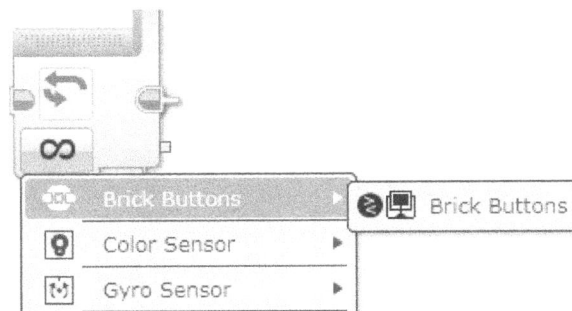

Red or Not

Modify your Red, Red, Red program so that it will say "red" when it sees red and "no" when it sees any other color. Add a switch to your program (under Flow) and set the button in the lower left corner of the Switch block to Color Sensor>Compare>Color. Set the color to 5, which corresponds to red.

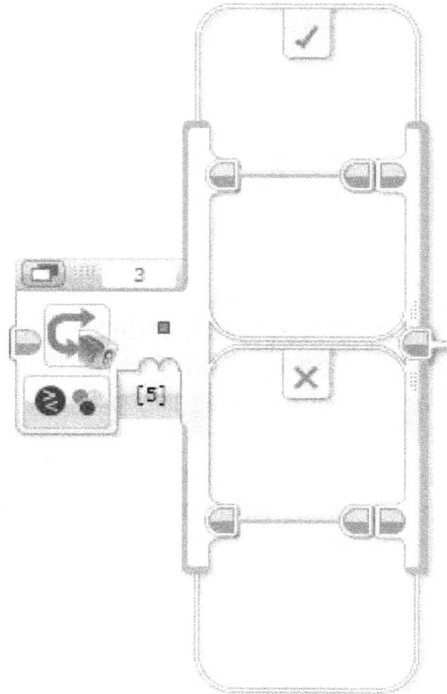

Hint: You will need to use a loop for this program in addition to the switch.

Rainbow Detector

To get really fancy, make a rainbow detector—one that can detect several different colors. Use a Switch block, with the button in the lower left corner set to Color Sensor—Measure—Color. To add more cases, click on the small plus button on the Switch block. You will need to designate one of the cases as the default. In this instance, use the No Color case as the default and leave that case empty. That way, if the sensor does not detect one of the colors on the list, the program will say nothing for that iteration of the loop.

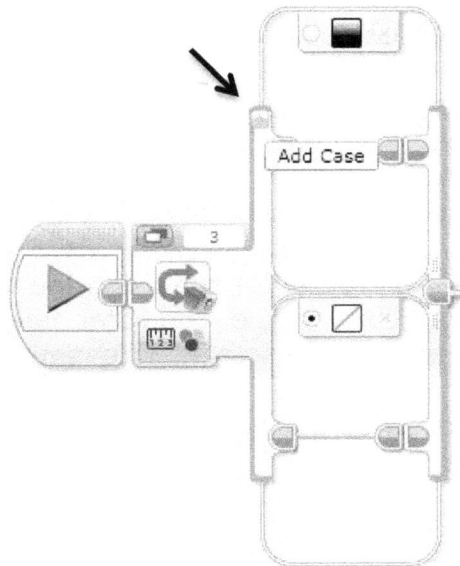

Teacher Information   Seeing Red

This sequence of activities introduces the basics of the EV3 Mindstorms software.

Objectives
1. See and Say: to learn how to use the color sensor and Wait block.
2. Red, Red, Red: to use a loop in a program.
3. Red or Not: to use a switch in a program.
4. Rainbow Detector: to use a switch with multiple cases.

Materials
EV3, color sensor, LEGO® pieces, computer

Time: Approximately 45 minutes

Notes
1. Before starting this sequence, you may want to show the students how to name and save a program.
2. Show the students that when a program is run using the Download and Run button with the EV3 still tethered to the computer, the execution of the program is highlighted on the computer screen. This feature is a valuable tool for debugging.
3. Meet and Greet is a good follow-up activity for students to work on after they finish Seeing Red. Students who finish Seeing Red quickly can design more elaborate greeter robots.

Sample Program for See and Say

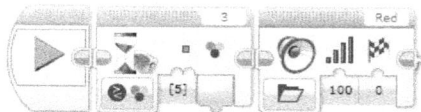

Sample Program for Red, Red, Red

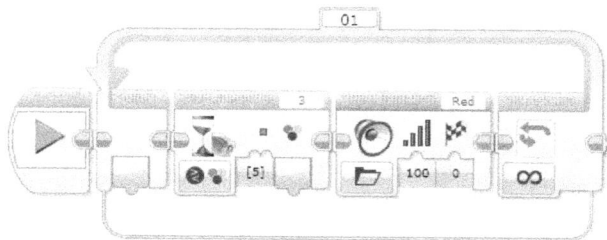

Sample Program for Red or Not

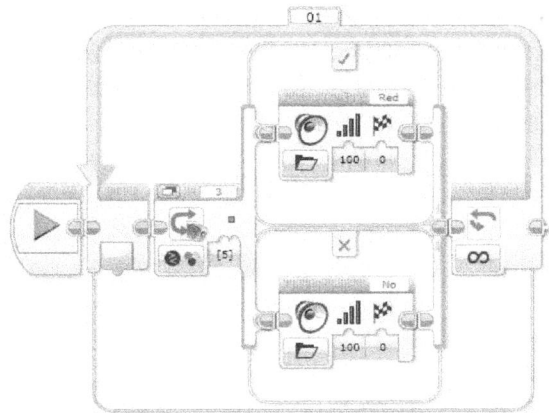

Sample Program for Rainbow Detector

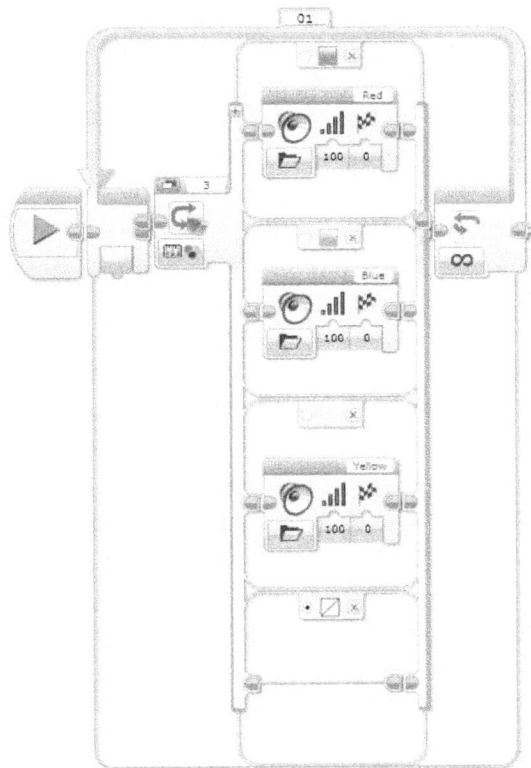

Driver's License

As you complete each programming activity, demonstrate your program to the teacher. Make sure that you get the stamp of approval for each activity before moving on to the next one.

Lurch	
Snake	
Cockroach	
Unsynchronized Motors	

Lurch

In this activity, you will write a program to make your car move with a lurching motion.

For this task, you will need to use a loop, a programming structure that repeats a sequence of steps over and over. In this case, your sequence will contain two parts, one to make the car drive forward for a short distance and one to make it stop for one second.

To make your car drive forward, you will use the rotation sensor that is built into the motor. On the Move Steering block, choose On for Rotations. Set the power to 100 (full power) and the number of rotations to two.

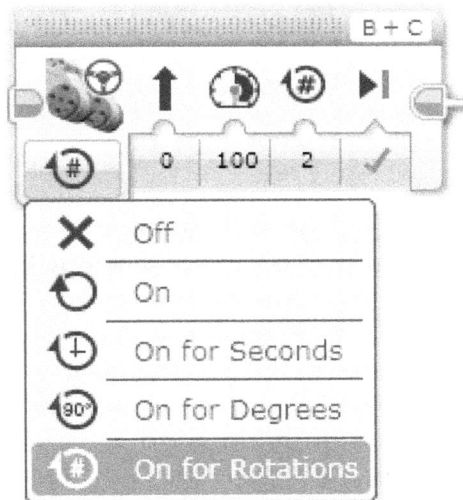

Snake

In this activity, you will write a program to make your car move like a snake, curving first in one direction and then in the other.

For this task, as in the previous one, you will need to use a loop. The sequence inside the loop will contain two parts, one to make the car curve to the left and one to make it curve to the right. However, you will use a different block to control the motors, the Move Tank block. While the Move Steering block adjusted both motors to make the robot turn, the Move Tank block lets you make the adjustments to each motor separately.

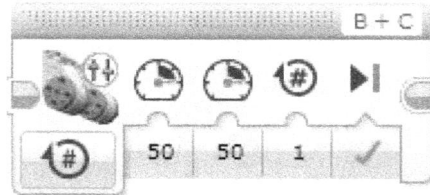

Once you have written your program, download it to the EV3 and run it. You will probably need to adjust the motor power and the duration to create a pleasing snake-like motion.

Cockroach

Program a cockroach. Cockroaches avoid light. The cockroach should crawl forward until a flashlight is shone on it. As soon as the cockroach detects the light, it should turn and scuttle away in the opposite direction. The behavior should be repeatable; each time your cockroach sees light, it should turn and scuttle.

To detect light, you will equip your cockroach with a color sensor. In addition to measuring color, the color sensor can be used to measure light intensity. The sensor measures light on a scale from 0 to 100, with brighter light producing higher numbers. The scale is arbitrary: it does not use standard measurement units such as lumens.

In your program, you will use a Wait block to wait for a brighter light. This can be done in two ways, by waiting for a particular light value or by waiting for a certain change in the light intensity. To use a particular value, choose Color Sensor>Compare. To wait for a change, choose Color Sensor>Change.

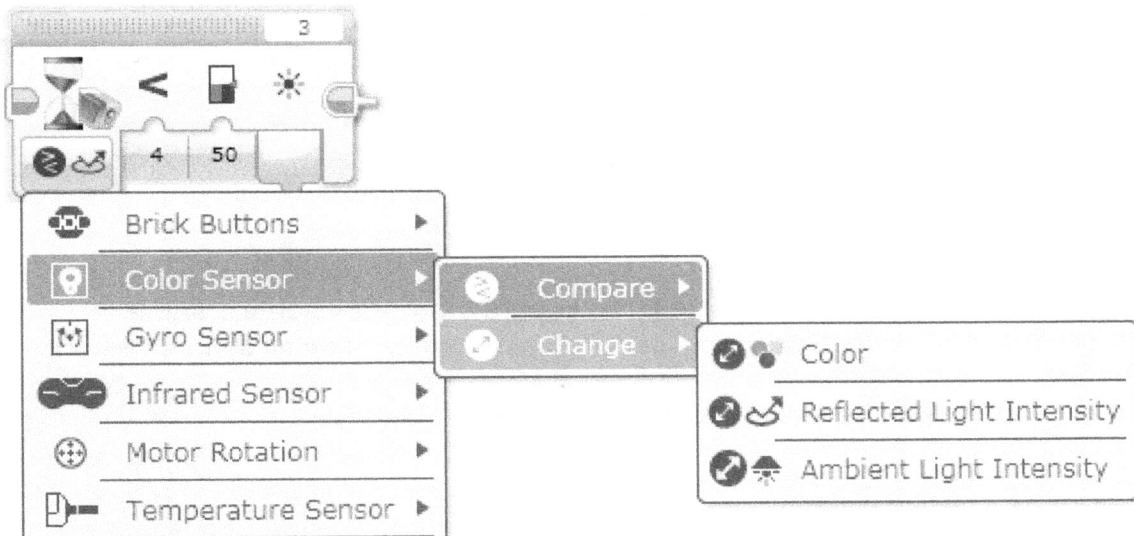

You can measure either ambient light or reflected light. Ambient light is the light in the room. If you select Reflected Light Intensity instead, the small red LED will turn on and the sensor will detect the light reflected from the LED as well as the ambient light. Experiment to see which one works better for the conditions your cockroach is facing.

Unsynchronized Motors

Write a program to run motors A and B separately. Motor A should alternate turning for one rotation and shutting off for one second. Motor B should alternate turning for two rotations and shutting off for one second. As a result, the two motors will be unsynchronized—sometimes one will be on, sometimes the other, sometimes both, and sometimes neither. Have the program end when a touch sensor is pressed.

Since the two motors are operating independently, they can be run in separate sequences. Each sequence will begin with its own Start block, which is located in the Flow Control palette.

Since the touch sensor to stop the program is independent of either motor, it can run in its own sequence as well. To stop the program, use the Stop Program block, which is found in the blue Advanced Palette.

The three sequences will look like three separate programs on the same Programming Canvas, but when you run the program, all three sequences will execute.

To get you started, here is the sequence to run motor A:

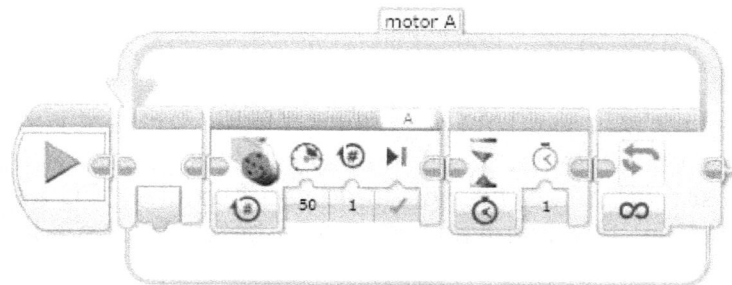

Teacher Information   Driver's License

These activities give students additional practice with programming and the light sensor.

Objectives
1. Lurch: to use a loop.
2. Snake: to use the Move Tank block for steering.
3. Cockroach: to gain practice using the color sensor.
4. Unsynchronized Motors: to write a program containing multiple Start blocks and the Stop block.

Materials
EV3, motors and leads, color sensor, LEGO® pieces, computer

Time: Approximately 50 minutes

Notes
1. Cloverleaf, Outside the Box, and Bug Battle are good follow-up activities for students to work on after they finish Driver's License.

Sample Program for Lurch

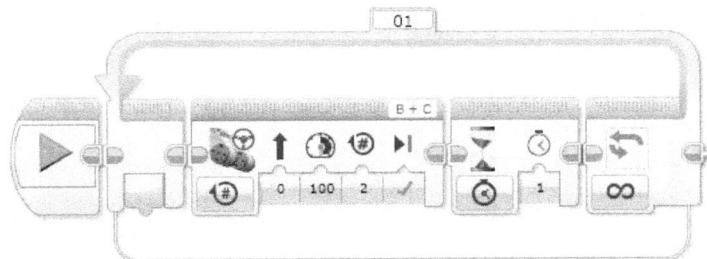

Sample Program for Snake

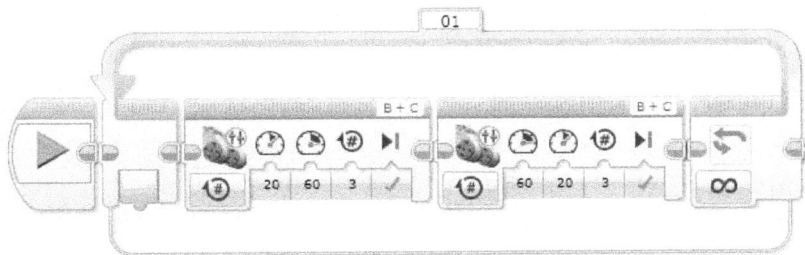

Sample Program for Cockroach

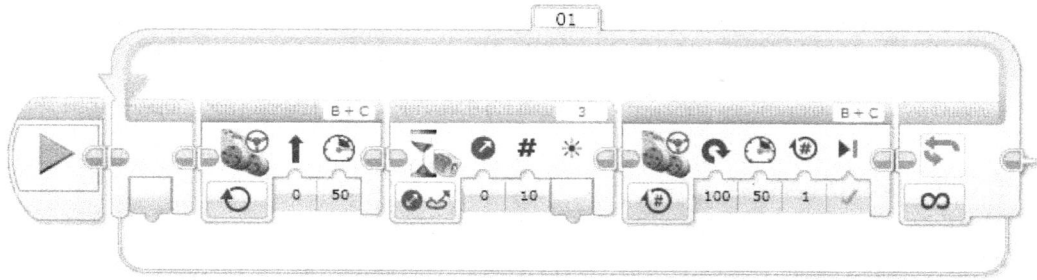

Sample Program for Unsynchronized Motors

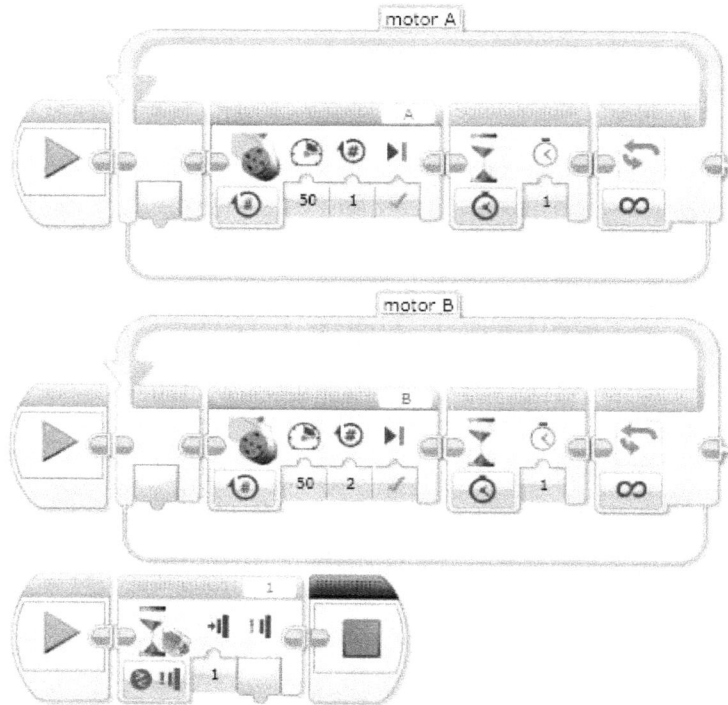

Fan Club

As you complete each programming activity, demonstrate your program to the teacher. Make sure that you get the stamp of approval for each activity before moving on to the next one.

Push-button Fan	
Daytime Fan	
Gradual Fan	
Five-speed Fan	

Push-button Fan

Build an EV3 fan. Use the medium motor, which can spin at a faster rate than the large motor, though it has less torque or turning force. The medium motor has a maximum RPM of around 250, as opposed to 170 for the large motor.

To run the medium motor, use the Medium Motor block:

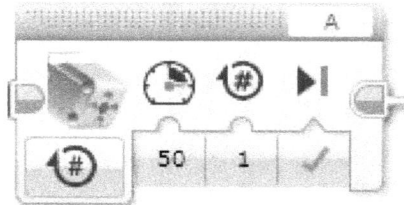

Program your fan to operate with a push button—the fan runs while the touch sensor is depressed and turns off while the touch sensor is released. For your program, you will need to use a touch-sensor-controlled switch, as well as a loop.

Since the motor will be controlled by the touch sensor, change the duration of the motor from On for Rotations to On.

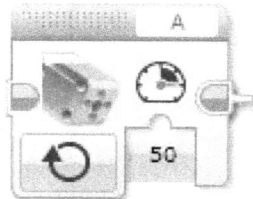

Daytime Fan

Program a daytime fan—one that goes faster as the light gets brighter and slows down when the light dims.

To program your fan, use a color sensor to measure the intensity of the light. Set the sensor to Measure>Ambient Light Intensity.

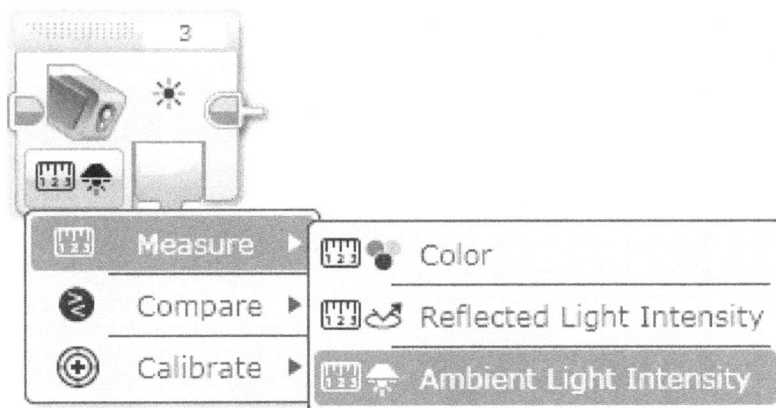

(Ambient light is the light in the room. If you select Reflected Light Intensity instead, the small red LED will turn on and the sensor will detect the light reflected from the LED as well as the ambient light.)

Use the output from the sensor as the input for the fan's motor, by connecting the two with a data wire.

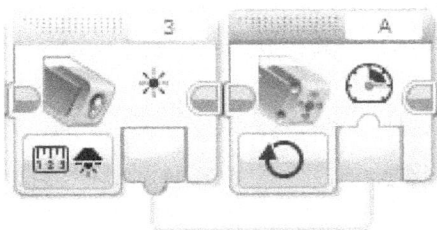

Data wires allow you to use the information from one block of a program to influence another block. To connect two blocks by a data wire, click and drag from the output of one block to the input of the other.

Gradual Fan

Program your fan to gradually increase its speed. That is, the longer it runs, the faster it goes. In order to track the increase, display the current speed on the EV3 screen for 0.2 seconds before increasing it.

For this program, you will use a data wire to control the motor speed. However, instead of connecting the data wire to a sensor, as you did in Daytime Fan, you will connect it to the loop index. The loop index keeps track of how many iterations of the loop have run. The first time through, the count is 0, the second time it is 1, and so on.

The loop below will increase the motor speed by one each time the loop runs.

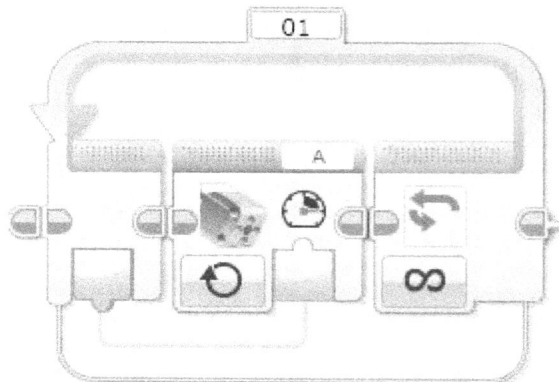

You will need to add more blocks to your loop to display the motor speed on the EV3 screen and to pause for 0.2 seconds before moving on to the next speed.

To display a value on the EV3 screen, use the Display block. Set Display to Text>Pixels and choose Wired in the box in the in the upper right of the block. Wire the value you wish to display to the Text node (the one marked T).

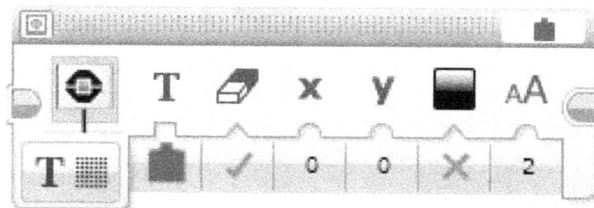

Five-speed Fan

Make a variable-speed fan. Give your fan six settings: off, very slow, slow, medium, high, and very high. Use the touch sensor to change the speed. Each time the touch sensor is pressed, the speed should change.

You will need to use nested loops for this program—one loop within another. The outer loop will run forever. The inner loop will run six times. You can use the loop index for this inner loop to control the motor speed. However, since the motor speed varies between zero and 100, you will need to multiply the loop counter by a constant before using it to run the motor.

The Math block, in the Data Operations palette, lets you manipulate a value from one block before sending it to another. In this case, you can set the block to multiply, wire the loop index into "a," type a multiplier in "b," and wire the result to the fan motor.

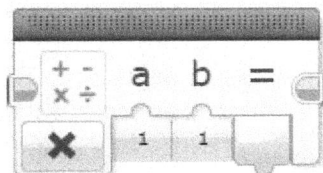

In your program, you will be using a Wait block with the touch sensor to control the speed. Set the State to 2 (Bumped) instead of 1 (Pressed). Can you figure out why? What might happen to your motor speeds if you used Pressed?

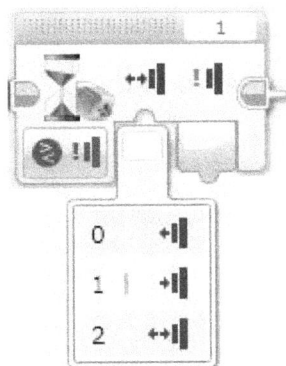

Teacher Information   Fan Club

This sequence of activities introduces the medium motor and more advanced programming topics, such as data wires and loop indices.

Objectives
1. Push-button Fan: to use the medium motor, touch sensor, and Switch block.
2. Daytime Fan: to use data wires in a program.
3. Gradual Fan: to use the loop index as an input.
4. Five-speed Fan: to use the Math block and nested loops.

Materials
EV3, touch sensor, color sensor, medium motor, LEGO® pieces, computer

Time: Approximately 50 minutes

Notes
1. In Five-Speed Fan, setting the touch sensor to Bumped rather than Pressed prevents a single push of the touch sensor from triggering multiple iterations of the loop.
2. Clean Sweep and Applause Meter are good follow-up activities for students to work on after they finish Fan Club.

Sample Program for Push-button Fan

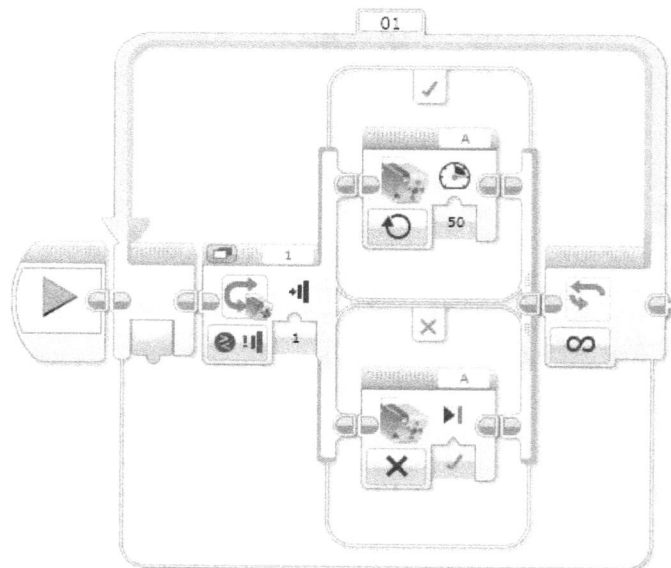

Sample Program for Daytime Fan

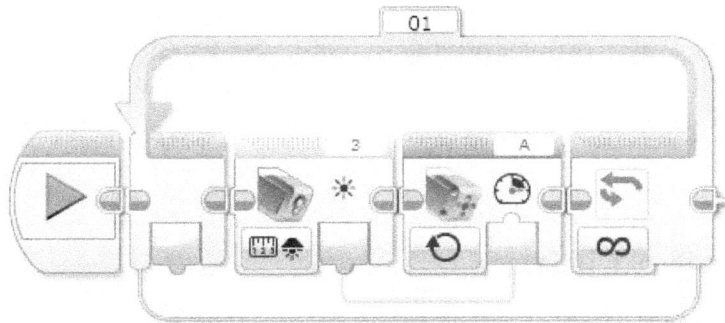

Sample Program for Gradual Fan

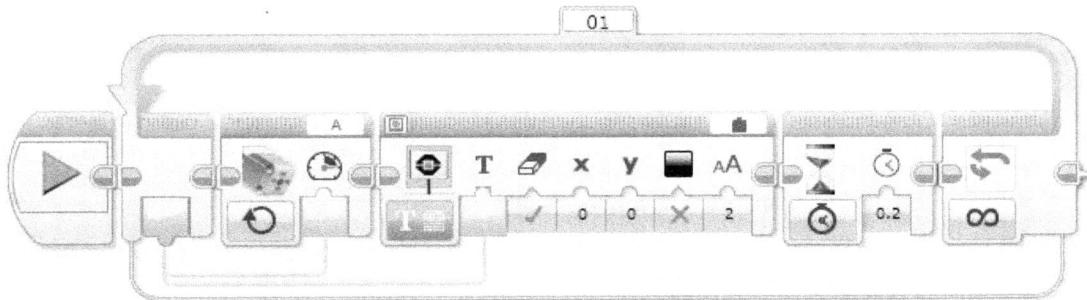

Sample Program for Five-speed Fan

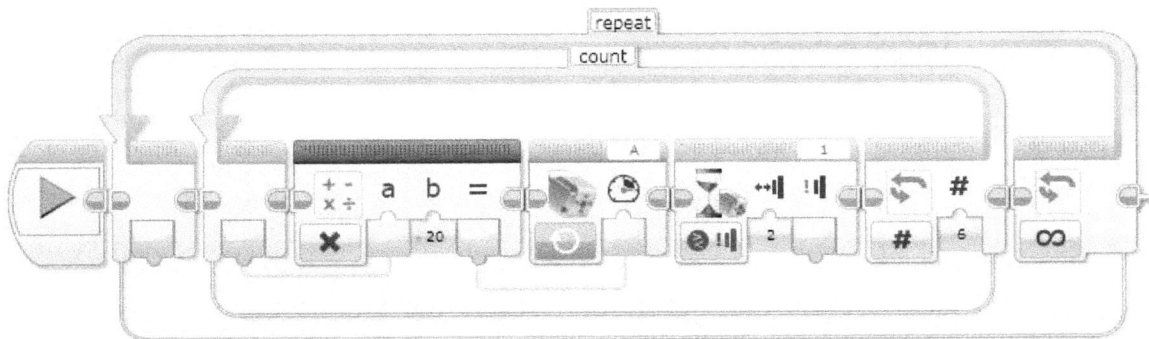

Traffic

As you complete each programming activity, demonstrate your program to the teacher. Make sure that you get the stamp of approval for each activity before moving on to the next one.

Traffic Light	
Walk Signal	
Animated Walker	
Intersection	

Traffic Light

Write a program to run a traffic light cycle—green, amber, red—over and over.
To create the lights, use the brick status lights of the EV3.

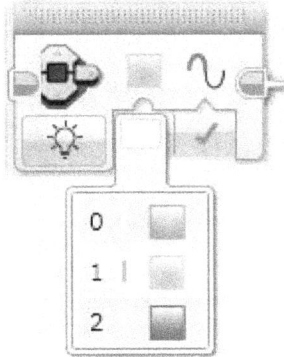

Set Pulse to False so that the light will be steady rather than blinking.

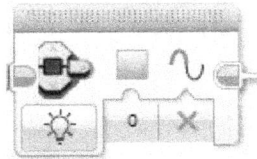

Run the red and green lights for three seconds each and the amber light for one
second.

Walk Signal

Write a program for a walk signal at the traffic light. Every time a touch-sensor button is pressed, a voice instructs the listener to walk and the EV3 screen displays "WALK" for three seconds.

To display "WALK," use the display block set to Text>Pixels. To adjust the position of the word, change the x and y coordinates.

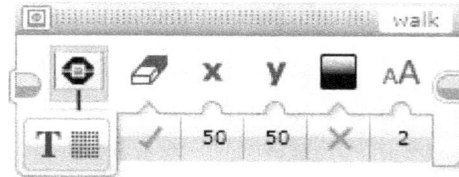

To make the WALK signal disappear after three seconds, use another display box, this time set to Reset Screen.

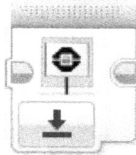

To add the voice, choose Sound Editor from the Tools menu. A pop-up box will appear. Click the red circle to begin recording and the blue square to stop. Once you are satisfied with your recording, save it. The saved sound will then appear under Project Sounds on the sound block. Set Play Type to 1: Play Once. That way, your program will start to play your sound and then move immediately to the next block.

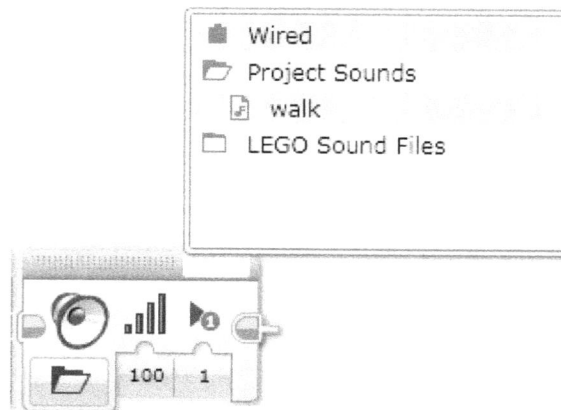

To make your walk signal run over and over, use a loop. You will need to decide upon the order of the blocks within the loop.

Animated Walker

Instead of having a walk signal that displays the word "WALK," make one with an animated walker. To do this, use the Image Editor, which creates display images, just as the Sound Editor creates sounds. The Image Editor is found in the Tools menu, just below the Sound Editor.

Open the Image Editor. You can use the drawing tools to create your own image. When you are satisfied with it, click on the image of the floppy disk in the upper left corner to save it. Once the image is saved, it is available to use in Project Images in the display block.

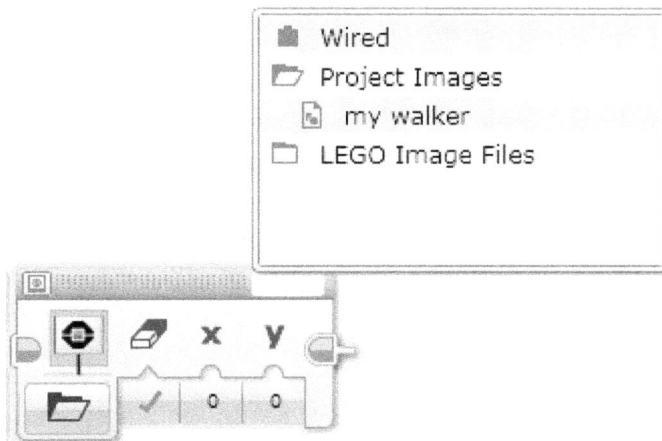

To make an animated figure, you will draw two similar images and alternate them repeatedly on the EV3 display. To create the two images, draw the first image, save it without closing the Image Editor, make changes to the image, then save it again under a different name.

In your program, put the two images in a loop, displaying one then the other repeatedly to produce the animation. (If you prefer to slow down the shift between images, add a Wait block after each image.)

Now, use nested loops to make the walk signal. When the touch sensor is pushed, your animated walker should play on the display for three seconds. After three seconds, the program should go back to waiting for the touch sensor to be pressed again.

Intersection

Combine the traffic light and the walk signal. Write a program that cycles the light through green, amber, and red until the WALK button is pushed. When the button is pushed, the light immediately turns red and the display indicates WALK for three seconds. At the end of three seconds, the WALK sign goes off and the traffic light returns to its usual cycle. If the WALK button is pushed again, the walk sequence should repeat.

To add the WALK signal, you will need to multitask, adding a second branch to your program. To do this, click on the wiring node of the Start block and drag your cursor from there to the first block of the new branch.

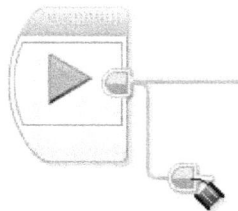

The new branch will wait for the touch sensor to be pushed, and then use a Loop Interrupt block (under Flow) to stop the green-amber-red sequence.

As a hint, your second branch will be very simple. It will look like this:

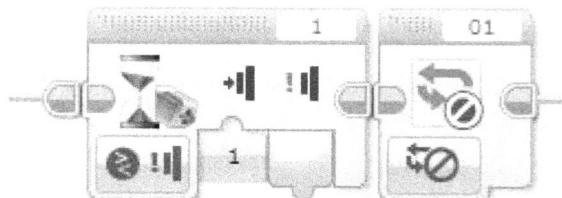

The first branch will be more complicated; it will contain both the traffic light sequence and the walk sequence.

You can choose which walk signal to use, either a simple text WALK or one of the fancier ones you created in Walk Signal and Animated Walker.

Teacher Information   Traffic

This sequence of activities introduces advanced features of the EV3 Mindstorms software, including the Sound and Image Editors, the brick status lights, multitasking, and the loop interrupt.

Objectives
1. Traffic Light: to use the brick status lights in a program.
2. Walk Signal: to learn how to use the Sound Editor and the display.
3. Animated Walker: to learn how to use the Image Editor and nested loops.
4. Intersection: to use multitasking and the Loop Interrupt block.

Materials
EV3, touch sensor, computer

Time: Approximately 60 minutes

Notes
1. For Intersection, the program can also be done with the walk-signal loop following the Loop Interrupt block on the second branch.
2. Tabletop Treasure is a good follow-up activity for students to work on after they finish Traffic.

Sample Program for Traffic Light

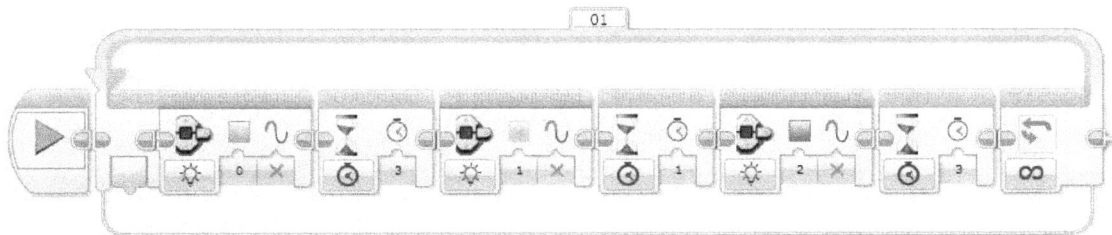

Sample Program for Walk Signal

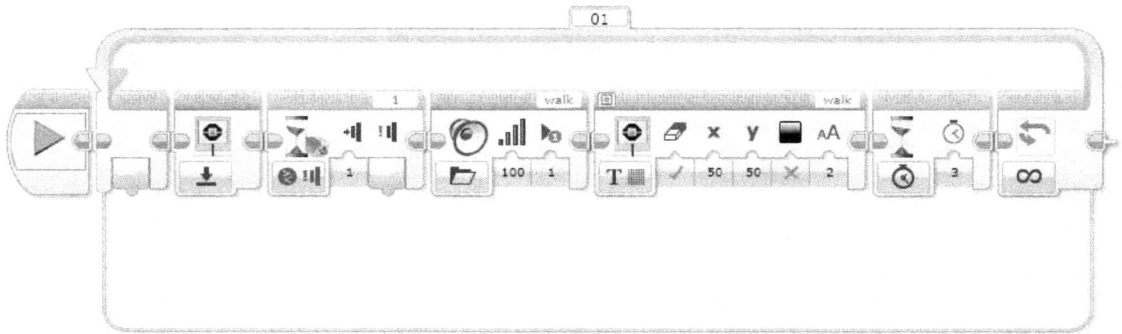

Sample Program for Animated Walker

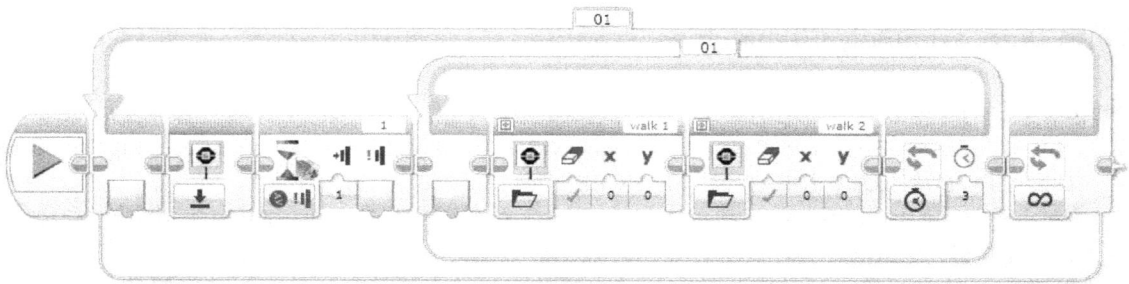

Sample Program for Intersection

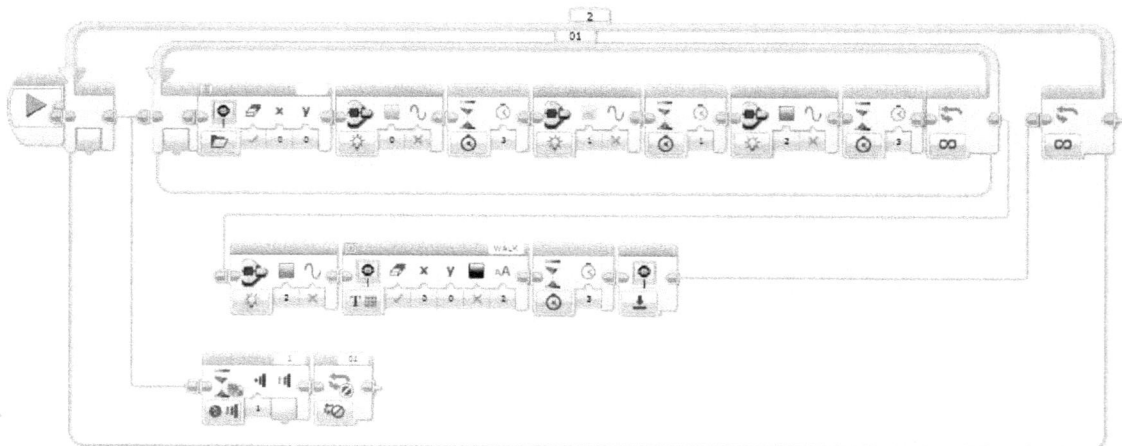

Number Sense

As you complete each programming activity, demonstrate your program to the teacher. Make sure that you get the stamp of approval for each activity before moving on to next one.

Roll of the Die	
Even or Odd	
Countdown	
Dog Year Converter	

Roll of the Die

The Random block, located in the red Data Operations palette, will generate a random number between a chosen minimum and maximum.

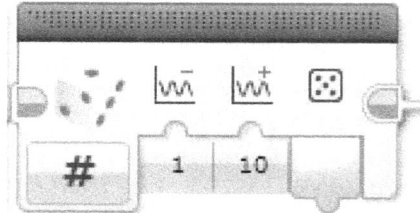

Use the Random block to write a program that displays a random number between one and six, like throwing a die. Each time the center EV3 button is pressed, a new number should be generated and displayed. To make it easier to tell that a new number has arrived, make the program play a sound to announce each new number.

Even or Odd

The program below displays the numbers from zero to ten on the EV3 screen at one-second intervals. The program uses the loop index at the start of the loop to generate the numbers, with the number increasing by one each time the loop executes. Note that the loop index is zero for the first run through the loop. The loop count at the end of the loop is set to 11 because the program will display a total of eleven numbers (zero through ten).

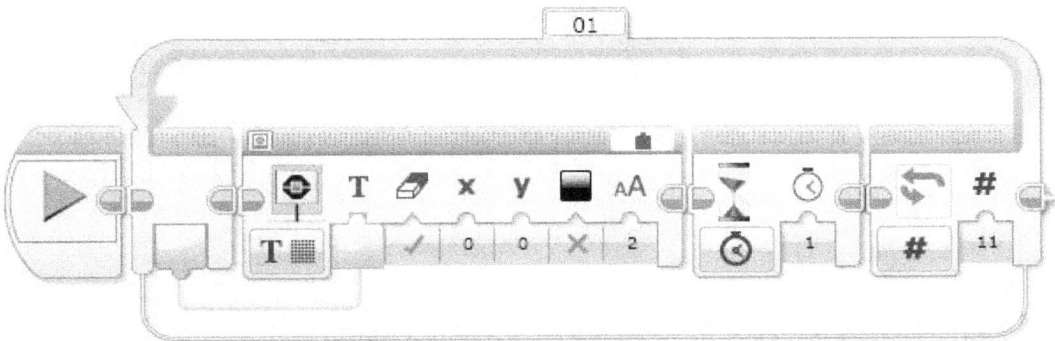

Write a program that counts by even or odd numbers (your choice), displaying each number on the EV3 screen for one second. That is, if you choose odd numbers, the screen should display 1, then 3, then 5, and so on. After displaying ten numbers, the program should end.

Remember that the loop index starts at zero, so you will have to figure out to make your display start with 1 (if you choose odd numbers) or 2 (if you choose even numbers).

Countdown

Take another look at the example program from Even or Odd, the one that counts from zero to ten.

How could you make it count in reverse?

Write a countdown program that counts down from ten to zero, displaying each number on the EV3 screen for one second. Hint: Use a Math block.

After the countdown reaches zero, the program should play a tone and then end.

Dog Year Converter

An often-used rule of thumb is that seven dog years are equal to one human year. That is, a two-year-old dog is actually fourteen years old in dog years.

Write a dog-year converter that will convert human years to dog years. To make your converter, use the random number generator to generate an age in human years. Every time you press a brick button, a new number should be generated. Your converter should calculate the equivalent age in dog years, then display the answer on the EV3 screen—the age in human years, the words "in dog years is," and the age in dog years. Thus, if the random number generated is 2, the screen should read, "2 in dog years is 14."

After you have converted the number, use the Text block to string together the original number, the text string " in dog years is " (with spaces), and the converted number.

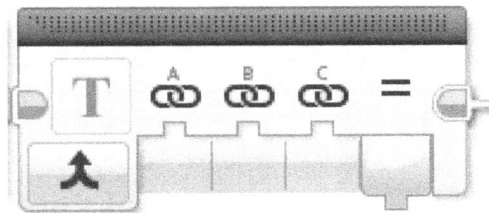

Wire the original number into A, the text string into B, and the converted number into C.

Teacher Information   Number Sense

This sequence introduces the Random block and the Text block. It gives additional practice using loops, data wires, and the Math and Display blocks.

Objectives
1. Roll of the Die: To learn how to use the Random block.
2. Even or Odd and Countdown: To gain additional practice performing calculations using the loop iteration number, the Math block, and data wires and to display the results on the EV3 screen.
3. Dog Year Converter: To learn how to use the Text block.

Materials
EV3, computer

Time: Approximately 60 minutes

Notes
1. Loaded Dice is a good follow-up activity for students to work on after they finish Number Sense.

Sample Program for Roll of the Die

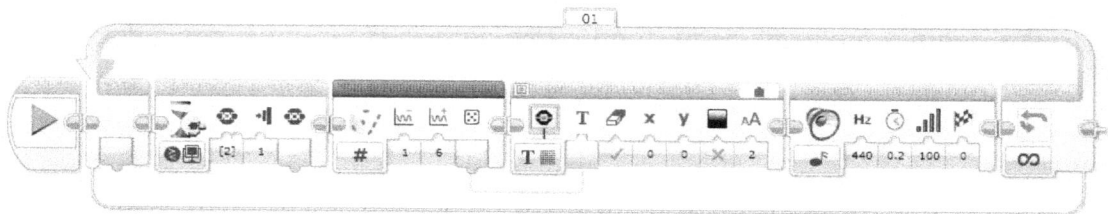

Sample Programs for Even or Odd

Even:

Odd:

Sample Program for Countdown

Sample Program for Dog Year Converter

Cautious Car: Logic Operations

Sometimes, a robot needs to respond to two different sensors.

Suppose you have a robot that is scared of things that go bump in the night. The robot drives forward unless it is dark <u>and</u> the robot bumps into something.

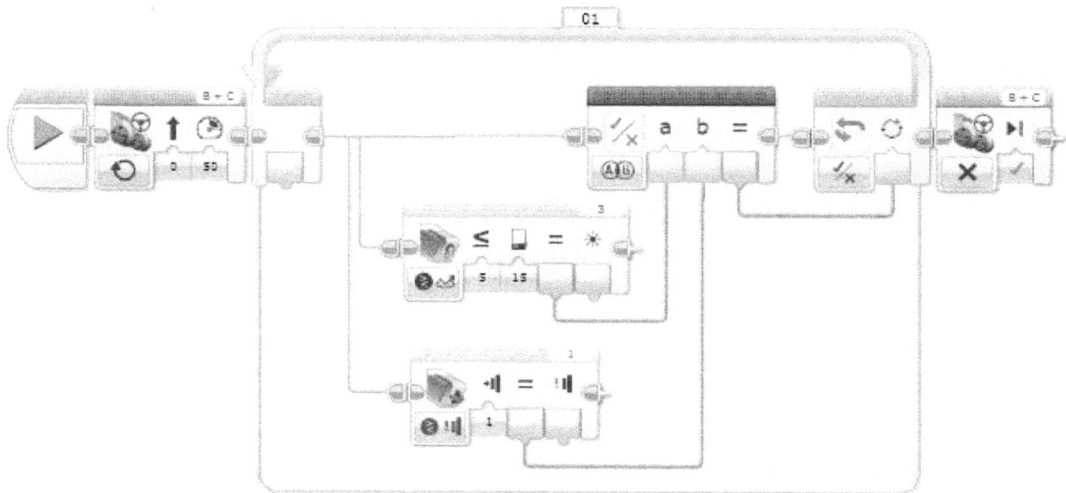

In the program above, the motors stay on as long as the loop is running. The loop ends when two conditions are met—the light sensor detects darkness and the touch sensor is pressed. The Logic Operations block, set to And mode, tracks these two conditions, returning a value of true when both conditions are met. At that point, the motors shut off.

Now, write a program for a cautious car—one that stops immediately if either its front OR rear bumper touches something.

Teacher Information   Cautious Car

This activity introduces the use of logic functions for controlling loops.

Objective
1. To learn how to use the Logic Operations block.

Materials
EV3 car, two touch sensors, computer

Time: Approximately 20 minutes

Notes
1. You may want to discuss logic functions before doing this activity.

Sample Program for Cautious Car

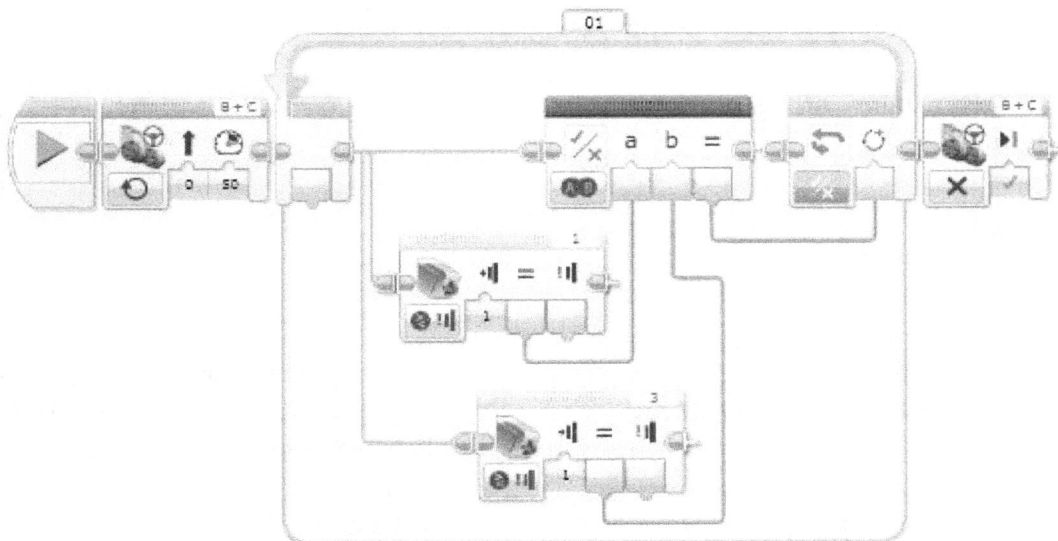

Touch Tally: Variables

Write a program to tally the number of times the touch sensor is pushed. After the tally is complete, the user presses the center brick button to signal the end of the tallying. The program then displays the result of the tally on the EV3 screen for five seconds.

For this program, you will need to use a variable. Here's how to create one: Choose the Variable block from the red Data Operations palette. Set the mode to Write - Numeric. Click on the box in the upper left corner to add a variable. You can also set the initial value for the variable.

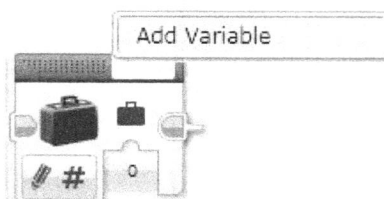

In addition to being used to create a variable, Write mode is used whenever you want to assign a new value to the variable.

To use the value of the variable in your program, you will need to switch it to Read mode.

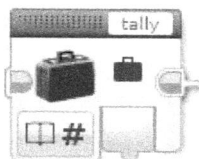

In your program, you will need to increase the value of the variable by one each time the touch sensor is pressed. To do this, wire the variable in Read mode to the Math block, add one, and wire the result to the variable in Write mode.

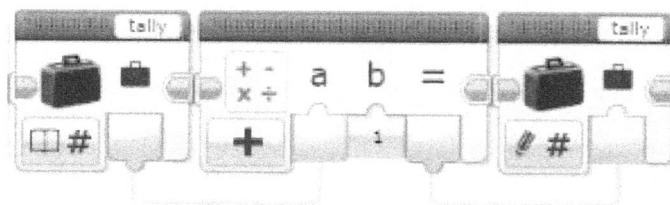

You will also need to use a Loop Interrupt block in a separate branch of the program to signal that the tally is complete.

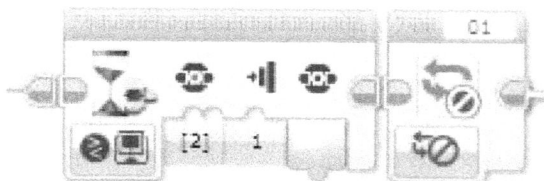

Teacher Information   Touch Tally

This activity introduces the use of variables.

Objectives
1. To learn how to use a variable in a program.
2. To practice the use of the Loop Interrupt block.

Materials
EV3 car, touch sensor, computer

Time: Approximately 20 minutes

Notes
1. You may want to discuss variables before doing this activity.

Sample Program for Touch Tally

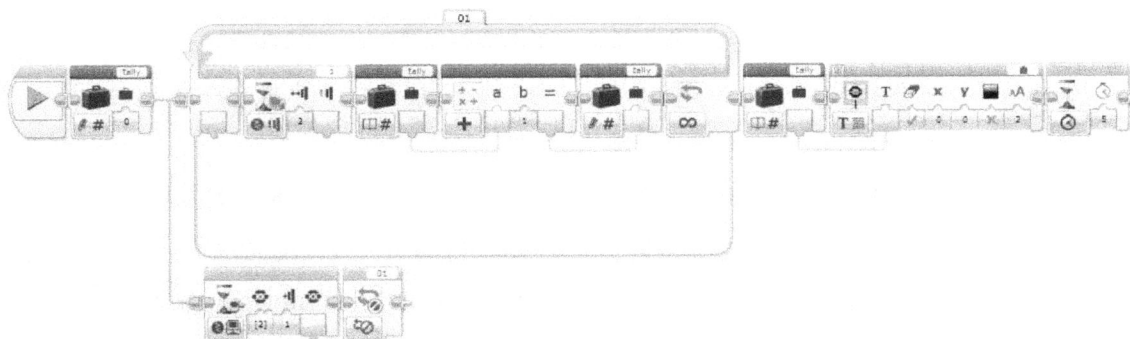

Brick-button Navigator: Arrays

The following program, Which Way, uses the brick buttons to decide which way to turn. The program waits for a brick button to be pressed. If the right button is pressed, the car drives forward for one second and then turns right. If the left button is pressed, the car drives forward and then turns left. The program uses the fact that the left brick button is assigned a value of one and the right button is assigned a value of three to know which way to turn.

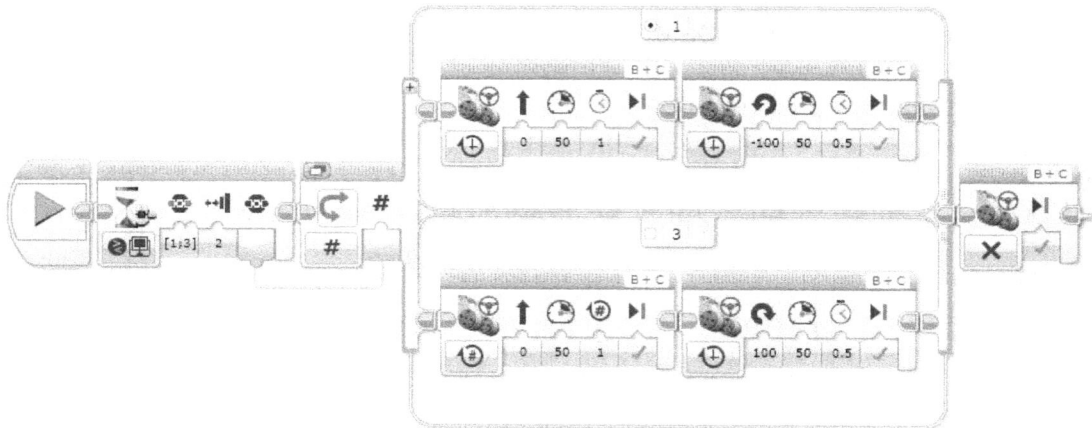

Write a fancier version of this program, one that lets you use the brick button to enter three turns at one time, which the car will then execute.

To do this, you will use an array keep track of the brick buttons pushed. An array lets you store a sequence of numbers.

You will use a loop like the one below to enter the button values into the array.

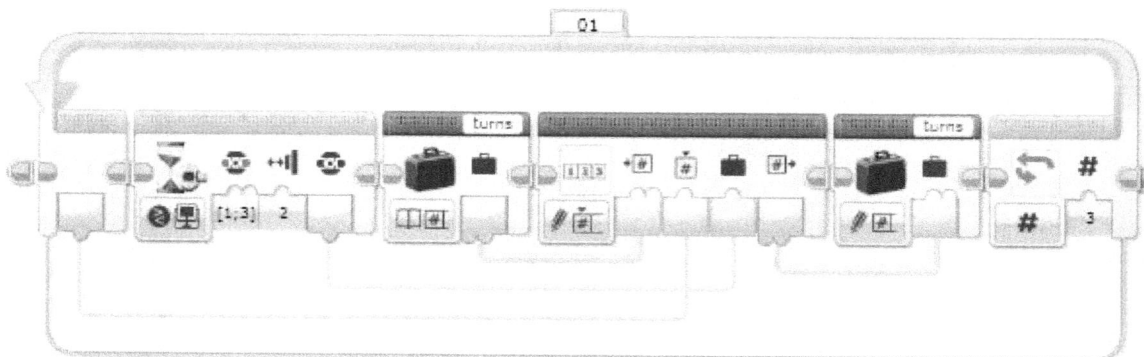

Let's look at the function of each block inside the loop. The first Wait block waits until a brick button is bumped and then sends that value to the array. The second Variable block, set to Read - Numeric Array mode accesses the array. The third Array block adds a new element, the latest button push, to the end of the array. The fourth block, another Variable block, this time set to Write - Numeric Array mode, receives the updated array from the Array block. The loop runs three times to register the three button pushes.

So, for your program, you will copy the loop above that populates the array and then write a second loop to run the car. The second loop will need to access an element of the array and use it to control the Which Way program above. The second loop will repeat three times, each time selecting the next element in the array and using it to turn the car.

This is a long and fairly complicated program. Take it slow and double-check your data wires. Good luck!

Teacher Information   Brick-button Navigator

The students learn how to use an array in a program.

Objectives
1. To learn how to use an array.

Materials
EV3 car, computer

Time
Approximately 30 minutes

Notes
1. This program is a complicated one. The students are given most of the program as a starting point to make the task less daunting.
2. You may want to discuss the example programs, Which Way and the array-building loop, as a class to make sure that the students understand them before they attempt their own programs.
3. You may want to go over array terminology, such as element and index, before doing this activity.

Sample Program for Brick-button Navigator

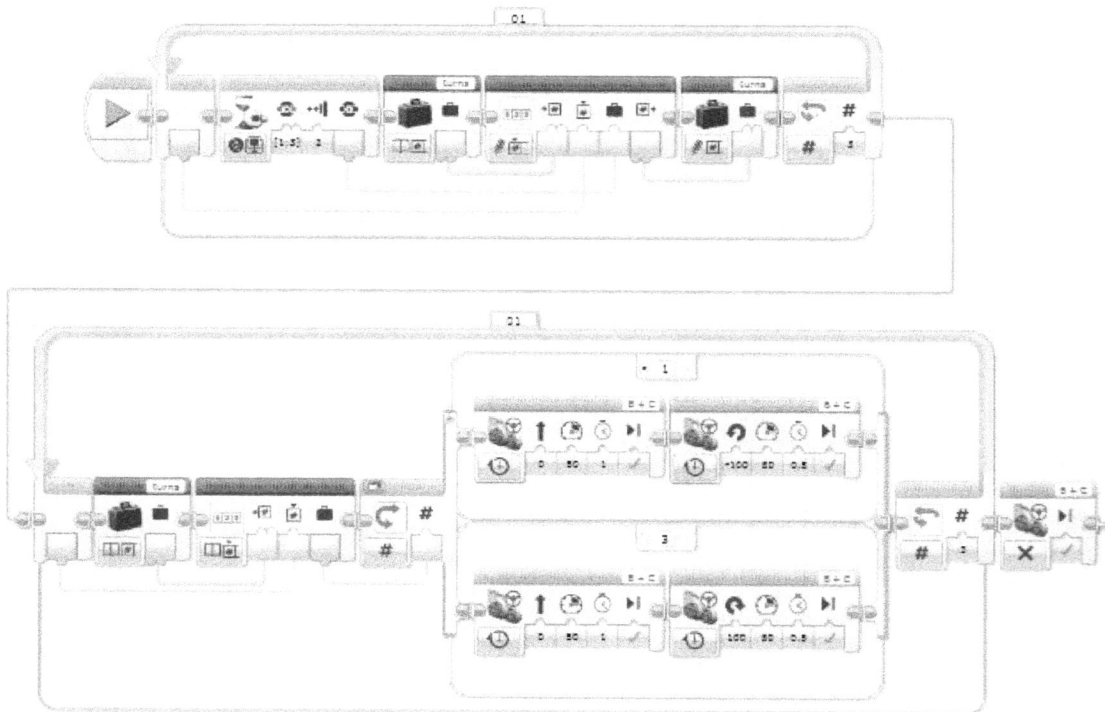

Mail Delivery: My Blocks

Sometimes you need to use the same sequence of code multiple times in a single program. Other times, your program may become so long and complicated that it is hard to follow. In either case, you may want to turn part of your program into a subroutine, which you can then represent with a single icon. Every time you want to use that sequence of code in your program, you can just place the subroutine icon in the program and the sequence will execute.

Subroutines in EV3 Mindstorms are called My Blocks. Suppose that you wanted to write a subroutine that would make an EV3 car spin for a specified number of seconds and then play a fanfare. Your code might look like this:

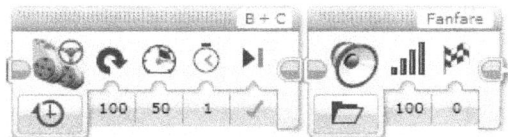

To turn the code into a My Block, first select both blocks (do not include the Start block). Then, click on Tools and choose My Block Builder.

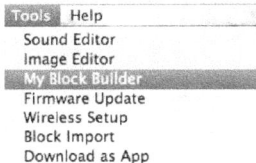

A pop-up window will open. Click on the Add Parameter button to add an input for the spin time. Use Parameter Setup to name your parameter and set its attributes. You can also choose a name for your My Block, write a description of it, and choose an icon for it.

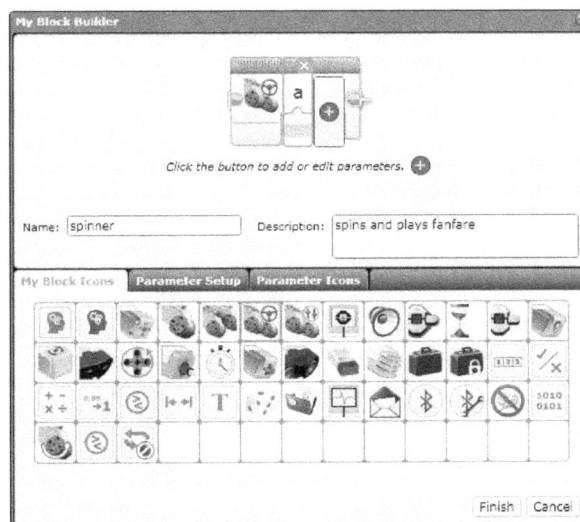

When you are done, click Finish. A new icon will appear on your screen, the parameter that you just created. Wire it to the appropriate place in your code. In this example, the parameter is wired to time on the Drive block.

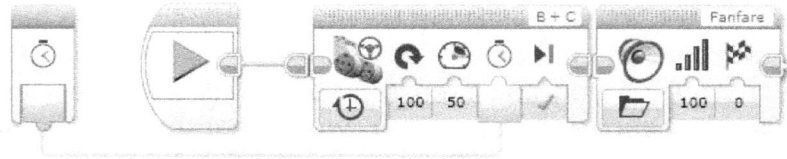

Your new block is now complete. It will appear in a new program window.

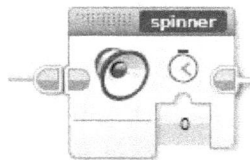

This new block can now be used in another program, just like any other block. It is housed in the aqua My Blocks palette. Double clicking on the new block will open it to show its component blocks.

Now that you've seen how My Blocks work, make and use one of your own. Write two programs. First, create a My Block, MailStop, which will drive forward for a specified number of seconds and then stop for two seconds. Second, create another program, Delivery Route, which will call MailStop three times. The EV3 postal van should make three stops, the first stop after driving for one second, the second after driving for an additional two seconds, and the third after driving an additional three seconds.

Teacher Information   Mail Delivery

This activity introduces the use of EV3 subroutines, called My Blocks.

Objectives
1. To learn how to create My Blocks.
2. To learn how to call My Blocks from another program.

Materials
EV3 car, computer

Time: Approximately 30 minutes

Notes
1. You may want to discuss subroutines before doing this activity.
2. Grassfire, in STEM Activities, makes use of My Blocks.

Sample Program for MailStop

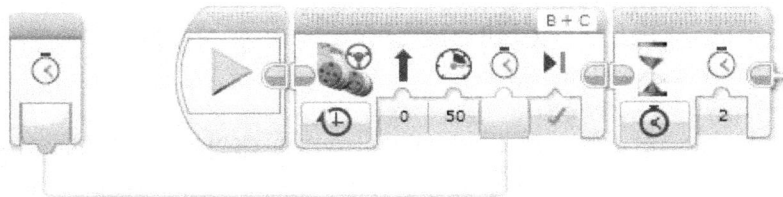

Sample Program for Delivery Route

Part Four: STEM Activities

These activities explore a range of STEM topics, especially ideas in physics, mathematics, and engineering design. The students gain a deeper understanding of the concepts by using them to solve problems. The concepts used in each activity are listed in parentheses below; see Appendix A for a complete listing by topic of all of the activities in this book.

The activities in this section are:
1. Getting Up to Speed (distance, graphing, variables, velocity)
2. Stop for Pedestrians (distance, velocity)
3. Parking Space (distance, graphing, velocity)
4. No Wheels (design, friction, velocity)
5. At a Snail's Pace (design, gears, rotation speed, velocity)
6. Cloverleaf (area)
7. Outside the Box (algorithms, navigation)
8. Bug Battle (design)
9. Tabletop Treasure (algorithms, navigation, trade-offs)
10. Spirographer (angles, parameters)
11. Puppy Bot (design, navigation)
12. Range Puppy (navigation)
13. Proportional Puppy (feedback control, navigation)
14. Haunted House (navigation)
15. Clean Sweep (design)
16. Perfect Pitcher (design, distance, levers, projectile motion)
17. Different Drummer (design, levers)
18. Ramp Up (center of gravity, design, gears, friction, torque, inclined planes)
19. Peak Performance (center of gravity, design, friction, torque, inclined planes, trade-offs)
20. Applause Meter (design, sound)
21. Musical Instrument (frequency, sound)
22. Random or Not (random numbers)
23. Loaded Dice (algorithms, probability, random numbers)
24. Efron's Dice (probability)
25. Voting Machine (data analysis, percentages)
26. Do You Have a Sister? (data analysis, percentages, probability)
27. Reaction Time (biology, variables)
28. Grassfire (navigation)

Getting Up to Speed

Build an EV3 car and see how fast it goes.

First, construct a car that can travel more than two meters in a straight line. Program it to drive forward for a little more than two meters. Choose a motor power that moves your car at a moderate speed—something that you can easily time with a stopwatch.

Place your car behind the starting line. Start the program and then start the stopwatch as the car crosses the starting line. Stop timing when the car crosses the finish line. Record your results in the data table below. Run two more trials. Find the speed for each trial, plus average time and speed for the three trials.

Trial #	Distance (m)	Time (s)	Speed (m/s)
1	1.0 m		
2	1.0 m		
3	1.0 m		
Average	1.0 m		

1. What is the advantage of running more than one trial in this experiment?

2. How long do you think it will take your car to run 0.5 meters? Make a prediction, and then test it.

   Predicted time:

   Actual time:

3. How long do you think it will take your car to run 2.0 meters? Make a prediction and test it.

   Predicted time:

   Actual time:

4. How long do you think it will take your car to run 1.25 meters? Make a prediction and test it.

   Predicted time:

   Actual time:

5. Now, make a time vs. distance line graph of your data, graphing your average for one meter, plus your data for the other three distances.

   Here are some things to remember about graphs:
   - Time (the independent variable) goes on the horizontal axis. Distance (the dependent variable) goes on the vertical axis.
   - The units on each axis must be evenly spaced (so, in this case, your data points will not be evenly spaced).
   - Each axis must be labeled and the graph must have a title.
   - Accuracy and neatness are essential; color and attractiveness are nice, but optional.

Teacher Information   Getting Up to Speed

In this activity, the students build simple cars and time them to find their speeds over various distances.

Objectives
1. To build a simple and reasonably sturdy car.
2. To calculate average speed from time and distance.
3. To use data to make predictions, then test them.
4. To make a line graph plotting time vs. distance.

Materials
EV3, motors, LEGO® parts including wheels, tape for marking lines, meter stick, stopwatch, computer

Time: Approximately 60 minutes

Notes
1. Before starting this activity, create a meter-long course for the cars with strips of tape for the start and finish. As the students do the activity, they will need to measure and add lines for 0.5 meters, 2.0 meters, and 1.25 meters.
2. After the students have graphed their data, discuss the significance of the graph with them. Most of the students should find that their data points lie more or less in a straight line, as the speeds of the cars are generally constant. (Because the first data point was not taken until after the car was already moving, the acceleration of the car at the beginning will not appear on the graphs.) Discuss the possible reasons for any deviation from a straight line, both the possibility that the car's speed was not constant and the possibility of experimental error.
3. If the students are familiar with the concept of slope, have them compare the slope of the line on their graph with the average speed they calculated for the car.
4. As an extension, have the students run their cars at different motor powers and plot the data sets on the same graph. The resulting graph provides a nice illustration of the relationship between speed and slope. For a further exploration of this topic, see Driving in Data Logging Activities.

Answers to Getting Up to Speed

Trial #	Distance (m)	Time (s)	Speed (m/s)
1	1.0 m	Answers will vary	Distance/Time
2	1.0 m	Answers will vary	Distance/Time
3	1.0 m	Answers will vary	Distance/Time
Average	1.0 m	Answers will vary	Average of the three speeds above

1. Taking the average of three runs usually improves the accuracy of your results by reducing the importance of any single error.
2. Predicted time: Answers will vary.
   Actual time: The time should be roughly half the time of the one-meter run.
3. Predicted time: Answers will vary.
   Actual time: The time should be roughly twice the time of the one-meter run.
4. Predicted time: Answers will vary.
   Actual time: The time should be roughly five-fourths the time of the one-meter run.
5. Graphs will vary. The line should be relatively straight, with its slope equal to the speed of the car.

Stop for Pedestrians

Pedestrians have the right of way. Can you program your car to stop for them?

A new pedestrian crosswalk is being planned for a two-meter-long road. Its exact location will be announced shortly. Before the crosswalk is constructed, you will be given some time to run your car for various distances along the road and collect data.

At the end of this practice time, the new pedestrian crosswalk will be built. You will be told exactly how far it is from the beginning of the road. You must program your car to drive as close as possible to the crosswalk without entering it. Once the crosswalk location is announced, you will not be allowed to run your car again. You may only program it.

For the test, you will place your car at the beginning of the road and, when given the signal, start the program. To successfully complete the challenge, your car must stop within ten centimeters of the crosswalk, but it must not knock over any minifigures in the crosswalk.

Good luck!

Teacher Information   Stop for Pedestrians

In this activity, the students run their cars over various distances and use the results to program the cars to go a specified distance.

Objectives
1. To collect data to solve a particular problem
2. To use data to make predictions, then test them.

Materials
EV3 car, tape for marking lines, meter stick, stopwatch, LEGO® minifigures, computer

Time: Approximately 40 minutes

Notes
1. This activity is similar to the following one; the teacher should choose the version that is most appropriate for the class. Stop for Pedestrians has a shorter roadway, encourages the students to make a graph but does not require it, contains an element of competition, and may well result in LEGO minifigures being knocked over. Parking Space features a longer and more challenging roadway, requires the students to make a data table and graph, and does not involve direct competition or potential injury to minifigures.
2. Before starting this activity, create a two-meter-long course for the cars with strips of tape for the start and finish.
3. Depending upon the students, you may want to give them some guidance in collecting data and encourage them to graph their results.
4. The crosswalk can be made from a long piece of tape stretched across the road. Make sure that it is parallel to the starting line. Populate the crosswalk with LEGO minifigures.
5. The assessment for this activity can be accomplished in different ways. One way is to require any students who are not successful to repeat the challenge, with different crosswalk locations, until they pass it. Another way is to note each group's results on the initial test and then give prizes—a small prize to any group which comes within ten centimeters of the crosswalk without going over; a slightly larger prize to the group that comes closest without going over.

Acknowledgement
This activity is based upon one by Professor Chris Rogers of Tufts University.

Parking Space

It is a long-standing tradition in some neighborhoods of Boston to clear the snow out of a parking space after a snowstorm and then save the space with traffic cones, lawn chairs, or other objects.

Such a parking space has been saved for you. All you have to do is park in it.

You will have time to practice driving on the three-meter-long street. You may use this time to try driving various distances and times. What you try is up to you; the only rule is that you must make a table and a graph of your data.

At the end of the practice time, you will be told exactly where your parking space is. After you are told, you may program your car, but you may not run it.

After all of the cars are programmed, we will run them one by one and see which ones are able to park entirely within the space.

Good luck!

Teacher Information   Parking Space

The students run their cars over various distances and use the results to program the cars to go a specified distance.

Objectives
1. To collect data to solve a particular problem
2. To use data to make predictions, then test them.
3. To interpolate from a line graph plotting time vs. distance.

Materials
EV3 car, tape for marking lines, meter stick, stopwatch, computer

Time: Approximately 40 minutes

Notes
1. This activity is similar to the previous one; the teacher should choose the version that is most appropriate for the class.
2. Before starting this activity, create a three-meter-long course for the cars with strips of tape for the start and finish.
3. Depending upon the students, you may want to give them some guidance in collecting data and graphing their results.
4. This activity can be made easier or harder by adjusting the size of the parking space. A thirty-centimeter-long space provides a challenge but will allow most students to be successful.

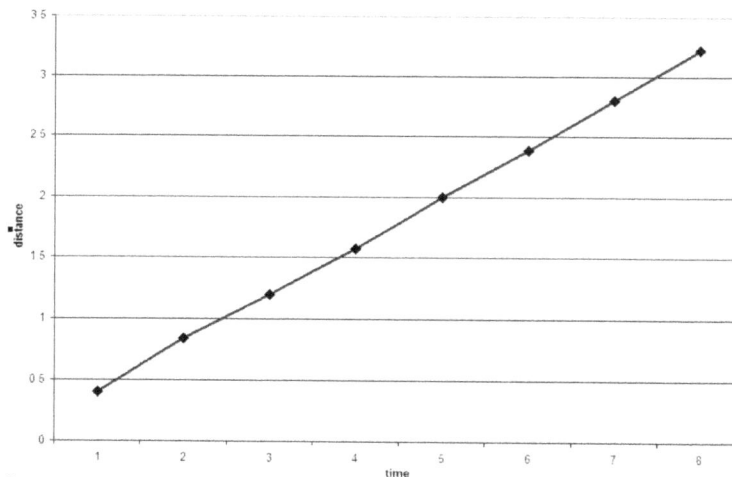

A graph of sample data.

No Wheels

Build an EV3 creature that moves. It may hop, squirm, crawl, whatever you like. However, it may not have wheels of any type.

1. Describe and sketch the mechanism you used to make your creature move.

2. Describe a design change you made as you were building and testing your creature. How did this change affect the creature's movement?

3. How fast does your creature move? Use a meter stick and a timer to calculate its speed in centimeters per second. Show your calculations.

Examples of Leg Mechanisms

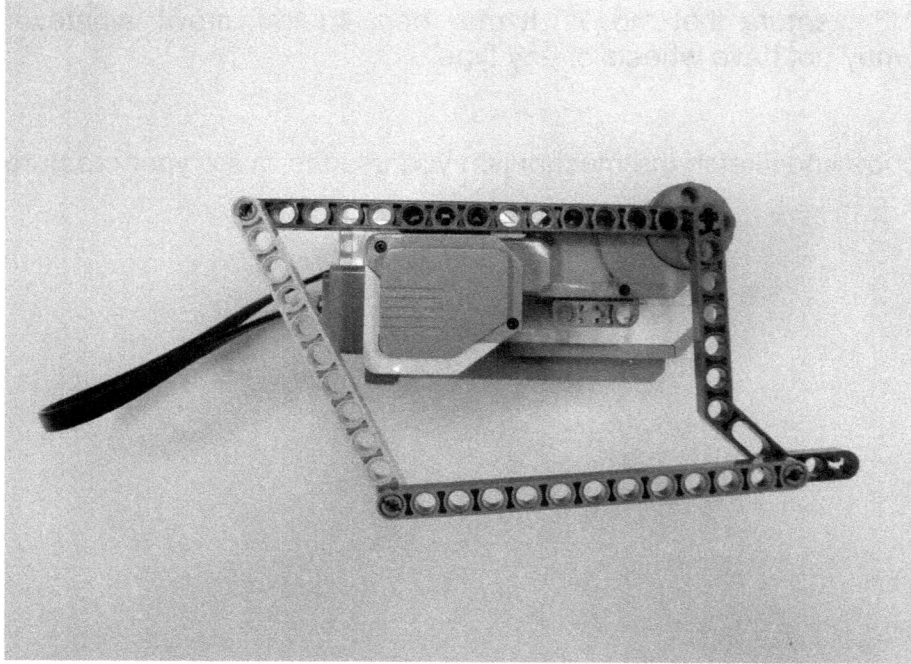

Four-bar linkage.

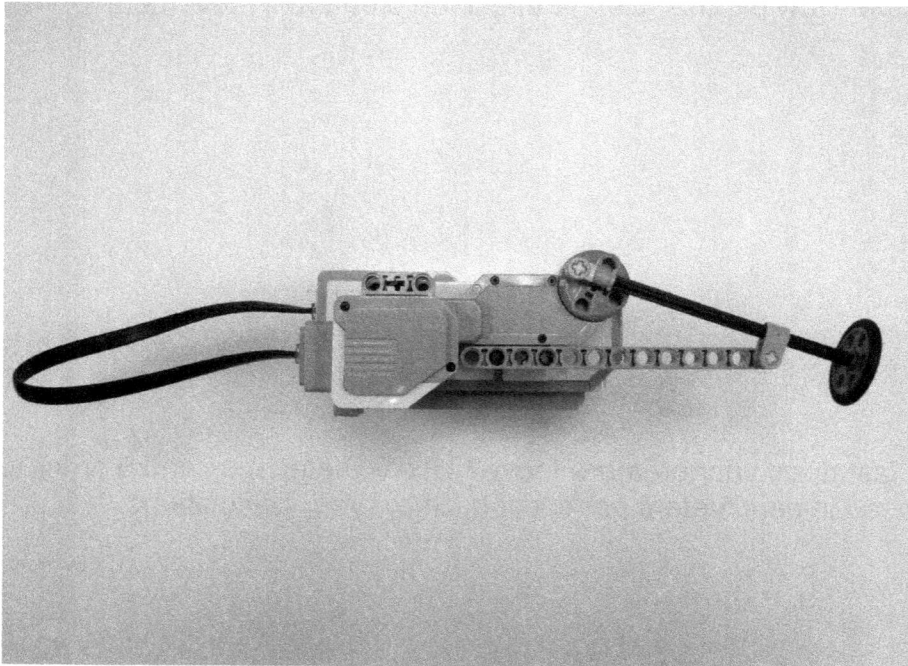

Reciprocating leg.

Teacher Information   No Wheels

The students design creatures that move without using wheels.

Objectives
1. To design a robotic creature that moves without wheels.
2. To use a meter stick and timer to calculate the speed of a robot.

Materials
EV3, motors, LEGO® pieces, meter stick, timer, computer

Time: Approximately 60 minutes

Notes
1. You may want to provide some examples of leg mechanisms to help the students get started. The Leg Mechanism page shows two possibilities.
2. Tell the students that they may use tires, though not as wheels. Tires positioned flat against the ground make excellent feet.
3. This activity provides an opportunity to discuss friction, as many of the initial foot designs will probably have trouble gaining enough traction to move the EV3 forward.
4. Encourage trial and error. After some experimentation, the students will begin to see what types of designs are successful in producing movement. You may want to have the students share some of their more successful designs with the rest of the class to expand the range of possibilities.
5. The motion of the creatures may be jerky and non-linear. You may want to discuss how this irregular motion complicates the calculation of the speed. Point out that the average speed may actually encompass a range of speeds.

Answers to No Wheels

1. Answers will vary. Encourage the students to use labels in the sketch.
2. Answers will vary.
3. Answers will vary. To calculate the speed, divide the distance traveled by the time taken to travel that distance.

At a Snail's Pace

Now that you know about velocity and gears, it is time for a snail race. As you know, snails are very slow. In fact, in this race, the slowest snail wins.

Here are the rules:

•Your snail must include an EV3, one or two motors, and one or more gear trains. You may also use craft materials, such as pipe cleaners and tissue paper, for decoration.

•Your snail must move forward at a constant pace during the race, though its forward motion can be very slow. In other words, no stopping, no turning, and no backing up.

•Your snail must look like a snail.

Once your snail is complete, document it:

•Calculate the slowdown—how many times your motor must turn in order for your wheels to turn once.

•Make a careful schematic of your gear train, showing how the gears mesh with one another. Your drawing does not need to be realistic; you may use symbols to represent the various sizes of gears. Be sure to include a key.

Good luck and may the slowest snail win!

Teacher Information   At a Snail's Pace

The students build snails for a snail race, where the slowest snail wins. This activity gives the students experience in building and analyzing complicated gear trains.

Objectives
1. To gain practice using gear trains, including compound gears and worm gears.
2. To be able to calculate gear ratios for complicated gear trains.
3. To design a gear train with its overall appearance and dimensions in mind (so that the result will look like a snail).

Materials
EV3, motors, LEGO® pieces including gears, craft materials, tape for starting line, computer

Time: Approximately 90 minutes

Notes
1. Before doing this activity, be sure that the students are introduced to compound gears and worm gears. The activities Gear Training and Worm Gears in the Low Tech Labs section are good preparation for this activity.
2. If you do not have access to a large supply of gears, you may want to set limits on how many gears each group may use.
3. Though this activity is billed as a race, try to de-emphasize the competitive aspects. In reality, many of the snails will be so slow that it will be nearly impossible to determine a winner.
4. You may want to add an additional constraint that the students must demonstrate that their snail moves forward, at least in theory, rather than moving backward (a real possibility, given the complexity of the gear trains).

Examples of Snail Schematics

Speedy the Snail

The motor must turn

3.66×10^{32} times

for the wheels to turn once.

Key

☐ = 40 tooth gear
▦ = 24 tooth gear
▪ = 8 tooth gear
▭ = Axle

[EV3] = EV3

☐ = Lego Block

▭ = 40-toothed gear
▭ = 24-toothed gear
▥ = 8-toothed gear

Snail Car Gear Train

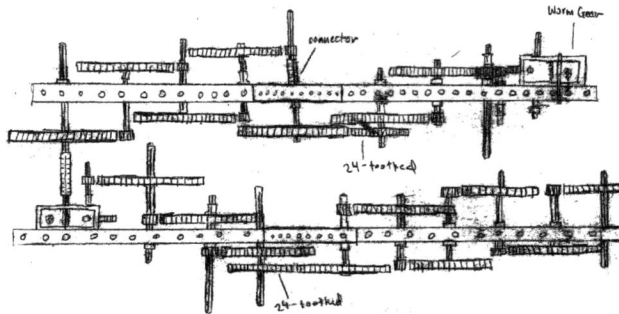

connector

Worm Gear

24-toothed

24-toothed

Speed: The motor must turn

9,765,624,992,000 times

for the tread to turn

JUST ONCE!

Cloverleaf

In this activity, you will program a car to drive in a cloverleaf pattern ⌘ and then calculate the area of the shape it produces.

To make the cloverleaf, you will need to program your car to drive straight and then turn in a circle to create one-fourth of the pattern. Use a loop to make your program repeat this straight/curve sequence four times. Write the program without worrying about the sharpness or timing of the turns. You will make the necessary adjustments after you have finished your initial program.

Once your initial program is complete, you are ready to fine-tune it. Attach a marker to your car so that you will be able to see the pattern your car produces clearly. Run your car and adjust the steering and timing in your program in order to perfect your cloverleaf.

Once you are satisfied with your cloverleaf, calculate its area. To do this, you will need to find the areas of the four circles and the area of the central square. (In doing your calculations, you can assume the shapes you drew are perfect circles and squares, even if the actual figures are not quite perfect.)

1. Find the area of one of the circles. Be sure to show your work.

2. Find the area of the central square. Show your work.

3. Find the total area. Again, show your work.

Teacher Information   Cloverleaf

The students write their own program to trace a cloverleaf pattern and then calculate the area enclosed by the figure.

Objectives
1. To gain practice using the Move Steering and Loop blocks.
2. To find the areas of circles and squares.

Materials
EV3 car, markers, large sheets of paper, computer

Time: Approximately 30 minutes

Notes
1. Once the students have written and tested the basic program, have them adjust the power, duration, and degree of turning to produce a cloverleaf. The motion of the cars will vary depending upon the floor surface, battery power, etc.
2. Some brands of thin markers fit nicely in the hole of the mini-turntable, which can be used as a penholder. Otherwise, students can build their own holders from LEGO pieces or tape the pens in place.

Sample Program for Cloverleaf

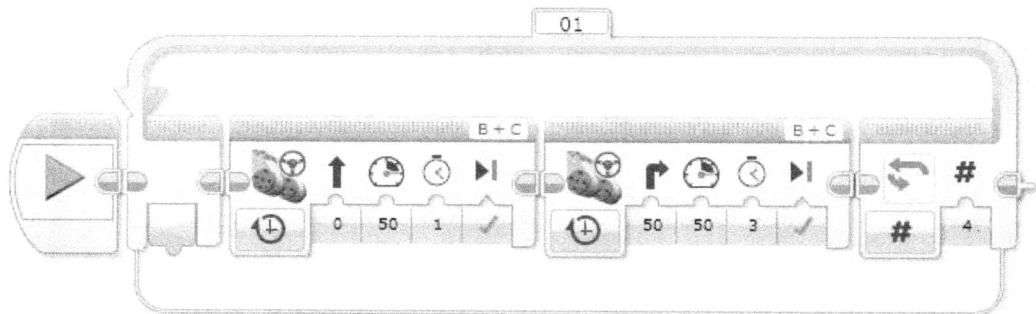

Answers to Cloverleaf

1. Answers will vary. The formula for the area of the circle is $A = \pi r^2$, where r is the radius of the circle.
2. Answers will vary. The formula for the area of a square is $A = s^2$, where s is the length of one side of the square.
3. Answers will vary. The total area of the cloverleaf will be four times the area of the circle plus the area of the central square: $A = 4\pi r^2 + s^2$.

Outside the Box

Build and program a robotic "bug" to escape from a box as quickly as possible.

The "box" will be a large rectangle with sides consisting of thick black lines. Your bug must find the gap in the black line and escape through it as quickly as possible. Your bug may stick part of its body over a line, but if it leaves the box completely by crossing the black line, it is disqualified.

A number of different algorithms can be used for this program. One simple approach is to have the bug move forward until its light sensor detects the black line, then reverse on a curved path for a short period, and then repeat this sequence using a loop.

A more sophisticated approach is to use a Switch block in the program, doing one behavior if the sensor detects light and a second behavior if the sensor detects dark.

Choose one of these algorithms and write an escape program for your bug. Test it and modify it if necessary.

Bug Battle

Now that your bug can escape the box, it's time for a competition. Two bugs at a time will compete to see which one is able to escape the box first. To start the contest, each team will place the other team's bug somewhere in the rectangle. Both bugs will be started at the same time.

Feel free to modify your hardware and/or software before the contest. Offensive and defensive tactics are allowed. Think outside the box!

Teacher Information   Outside the Box/Bug Battle

The students build bugs that escape from a box as quickly as possible.

Objectives
1. To write a program that uses a light sensor.
2. To create a vehicle that turns easily.
3. To solve an open-ended problem.
4. To test and modify a design repeatedly.

Materials
EV3, motors, light sensor, LEGO® pieces including wheels, white paper and black tape to make the box, computer

Time: Approximately 30 minutes for each activity

Notes
1. To build the box, make a large square or rectangle with thick black tape on a white background. Leave a thirty-centimeter gap in the tape on one side to create the opening. Of course, you can adjust the size of the box and of the opening to make the activity harder or easier.
2. Let the students try their bug in the box as often as they wish. The process of testing and redesigning is an essential part of this activity.
3. The light sensor works best if it is placed close to the ground and facing the floor.

Sample Wait-block Program for Outside the Box/Bug Batttle

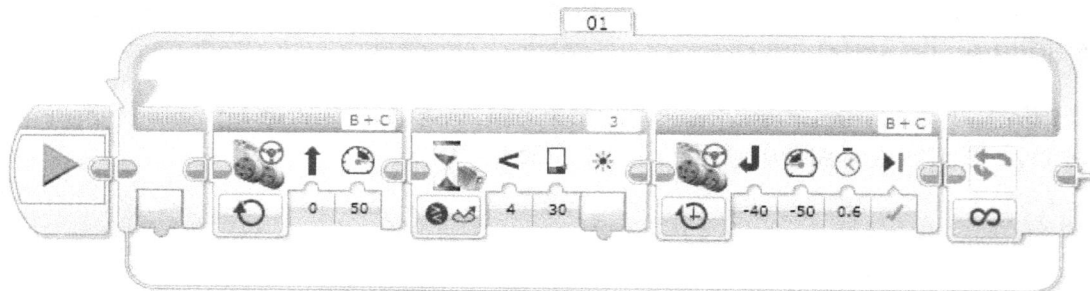

Sample Switch-block Program for Outside the Box/Bug Battle

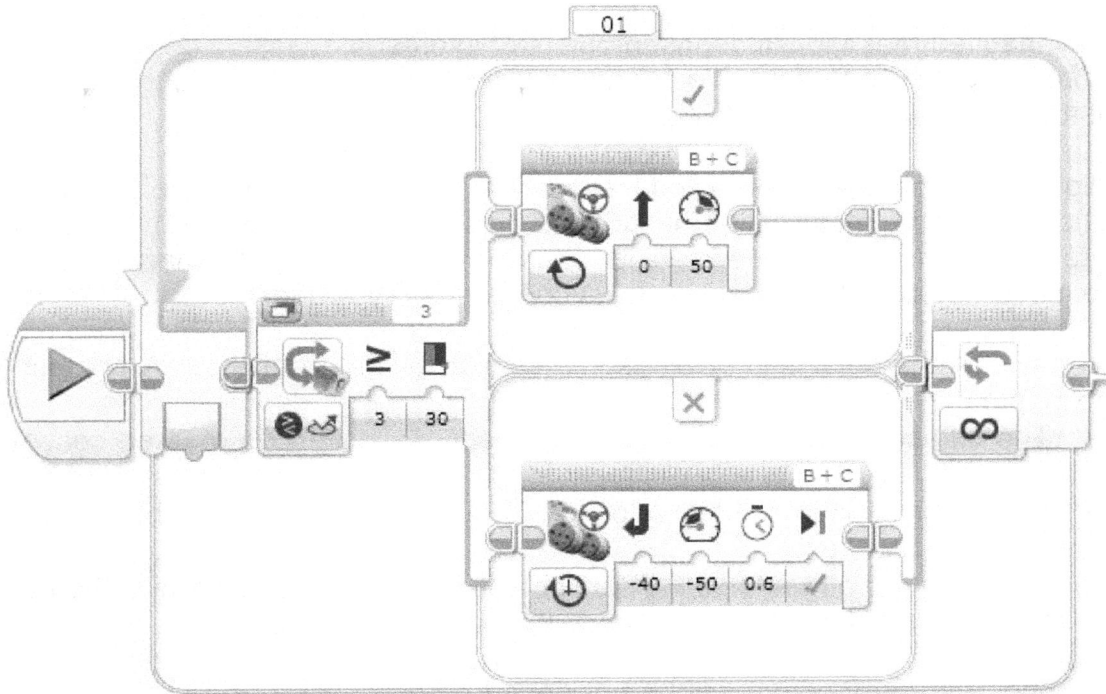

Tabletop Treasure

An uninhabited mesa is suddenly an object of interest, because a golden treasure is rumored to lie on its surface. You (and other treasure seekers) are sending robots to look for the treasure. Who will find it first?

Design and program a robot to find the treasure as quickly as possible, without falling off the edge of the mesa. The treasure is a gold spot (detectable by the color sensor set to yellow), located somewhere on a tabletop. Once it finds the treasure, your robot should stop on the spot and play a congratulatory sound.

You may use the color and/or ultrasonic sensor for this challenge. A few things to keep in mind:

- The ultrasonic sensor will not work reliably for distances of three centimeters or less.
- The color sensor detects color best if it is placed close to the surface and perpendicular to it.
- Mounting the edge-detecting sensor well in front of the car allows the car time to respond before its wheels reach the edge of the table.

Your robot and the treasure will be positioned on the table at the start of your turn. You will not know in advance where they will be placed. The robot that is able to locate the treasure in the shortest amount of time is the winner. Robots that drive off the table are disqualified.

1. A number of different algorithms are possible for solving this challenge. Describe the one you used.

2. Often, engineering projects involve trade-offs between competing goals. Describe a trade-off that you had to make in the activity.

Teacher Information   Tabletop Treasure

The students run their cars on a tabletop, trying to locate the golden treasure as quickly as possible without driving off the edge of the table.

Objectives
1. To devise and test an algorithm for solving a particular problem.
2. To weigh trade-offs in solving a problem.
3. Depending upon the algorithm chosen, to use the Loop Interrupt block or a multi-case Switch block in a program.

Materials
EV3 car, ultrasonic sensor, color sensor, (preferably low) table, yellow paper for treasure, stopwatch, computer

Time: Approximately 50 minutes

Notes
1. The "golden treasure" can be a square of yellow paper. Test the paper ahead of time to make sure that the color sensor identifies it as yellow reliably.
2. A low table is desirable for this activity, for obvious reasons. You may want to place padding underneath the table and/or place spotters around the table to catch cars if they drive off the edge.
3. This activity is similar to (though more complicated than) Outside the Box, which would make an interesting comparison to this activity—in Outside the Box, the robot is searching for an opening on the perimeter of the box. In this activity, the robot is searching for a small object in the interior of the box.

Sample Program for Tabletop Treasure Using Two Sensors

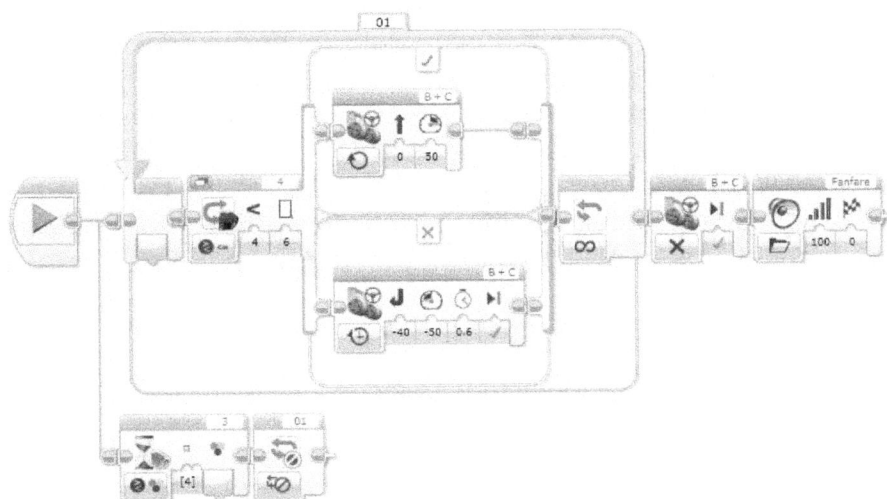

Sample Program for Tabletop Treasure Using One Sensor

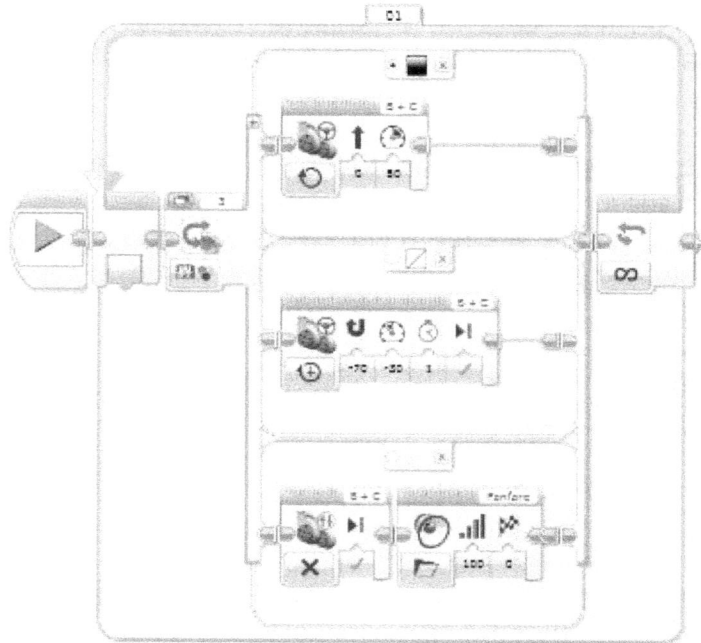

This program is written for a black table. The colors can be adjusted for other tables.

Answers to Tabletop Treasure

1. Answers will vary. One possible algorithm is to use the ultrasonic sensor to detect the edge of the table, backing away and turning as in Outside the Box. At the same time, the color sensor is used to detect the gold paper and the car is halted. Another possibility is to conduct a more systematic back-and-forth search of the area, using the ultrasonic sensor to avoid the edge and the color sensor to detect the gold paper. A third possibility is to use a multi-case switch, driving forward if the color sensor detects the tabletop color, stopping if it detects yellow, and backing and turning if it detects any other color.

2. A number of answers are possible. One common trade-off is speed and reliability. Going faster lets the car cover the field more quickly, but increases the chances that it will inadvertently drive off the table.

Spirographer

Spirograph is a children's toy that uses gears and rings with teeth to produce geometric designs. Use an EV3 robot with a gyro sensor to make your own spirographic designs.

Build a two-motor EV3 car and attach a gyro sensor to it. Attach one or more pens to the car. If you are using thin markers, the mini turntable can be used as a penholder.

To draw a design, your robot needs to drive forward, turn using the gyro sensor, and then repeat these two steps over and over. To execute the turn, use the Wait for Gyro block, under Flow, setting it to Compare, then Angle.

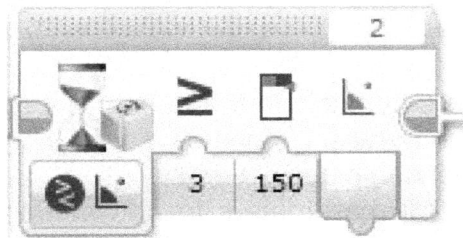

Since your program will call the gyro sensor repeatedly, you will need to initialize the sensor before it is used each time. To do this, go to Sensor, then Gyro Sensor. Choose Reset Gyro.

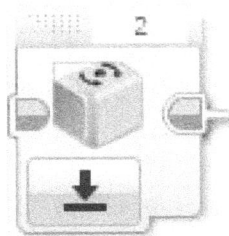

Once your basic spirographer is working, try making adjustments. How can you change the radius of the design? The number of petals of your flower or points of your star? What other changes can you make?

Make at least three different designs.

Choose one of your designs to document. Print out the program that created the design and attach it to the design.

Teacher Information   Spirographer

The students use the gyro sensor to produce spirograph-like designs, varying parameters to see how the design is affected.

Objectives
1. To gain additional practice using the gyro sensor.
2. To see how varying parameters will affect the final outcome.

Materials
EV3 car, gyro sensor, markers, large sheets of paper, tape, computer

Time: Approximately 50 minutes

Notes
1. Some brands of thin markers fit nicely in the hole of the mini-turntable, which can be used as a penholder. Otherwise, students can build their own holders from LEGO pieces or tape the pens in place.
2. You may want to structure the parameters for the students, assigning particular ones for them to vary. Ones to try include radius of the circle, placement of the pen(s), angle of the gyro sensor, sharpness of the turn, and duration of the turn.

Sample Program for Spirographer

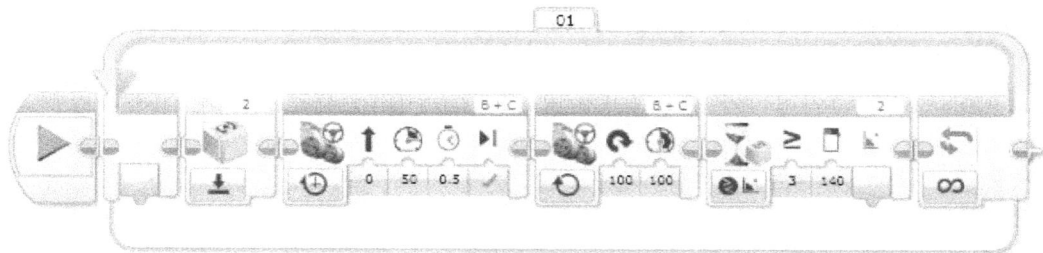

Puppy Bot

Make a puppy that will stay close to your heels. If you get further away, it will speed up to reach you. If you slow down or stop, it will too.

To program your puppy bot, use an ultrasonic sensor, which measures distance. Use the output from the sensor as the input for the puppy bot's motors, by connecting the two with a data wire.

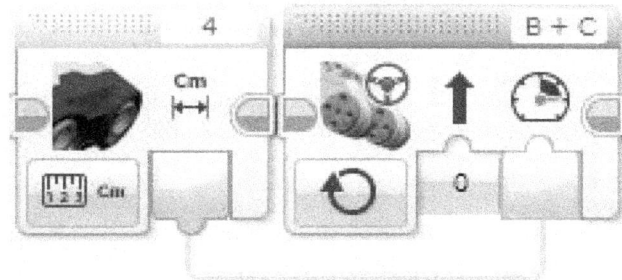

Depending upon the orientation of your motors, you may find that your puppy goes backward. How can you adjust your program to make the puppy go forward instead? Hint: Use a Math block.

Once you have trained your puppy to follow you, give it some personality. Add a head that turns or a tail that wags.

1. Which feature did you add?

2. How did you change your computer program to run the additional motor?

The EV3 sound files include a number of dog noises. As a finishing touch, give your puppy a voice.

Teacher Information   Puppy Bot

The students create a puppy that will follow them.

Objectives
1. To gain additional practice using a sensor to control a motor.
2. To write a program to perform two independent actions at the same time.

Materials
EV3, motors, ultrasonic sensor, LEGO® pieces including wheels, computer, cardboard (optional)

Time: Approximately 50 minutes

Notes
1. This activity is designed to be done as the first in a sequence of activities: Puppy Bot, Range Puppy, Proportional Puppy, and Data Logging Puppy. It can also be used as a stand-alone activity.
2. Since the ultrasonic sensor works best when triggered by a flat surface, you may want to make large cardboard "dog treats" to use in guiding the puppies.
3. If the puppy goes backward, the students will need to insert a Math block into their program to multiply the output of the ultrasonic sensor by negative one before feeding it into the motor.

Sample Puppy Bot Program with Separate Start Blocks

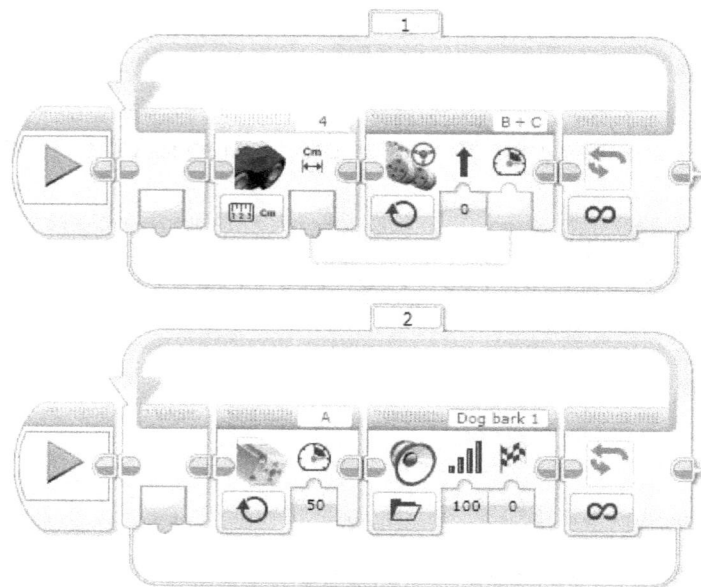

Sample Puppy Bot Program with Multitasking

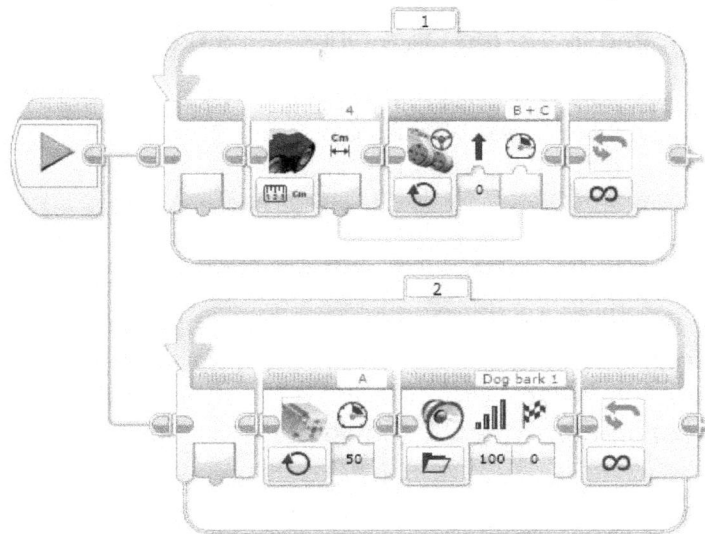

Answers to Puppy Bot

1. Answers will vary. Common features are a wagging tail, flopping ears, or a turning head.
2. Answers will vary. There are a number of ways to add movement and barking to the original program. It can be done using multitasking or using separate start blocks. See the sample programs above for an example of each approach.

Range Puppy

Program your puppy to follow you at a distance of twenty centimeters. To do this, use the Range block, located in the red Data Operations palette, set to inside mode.

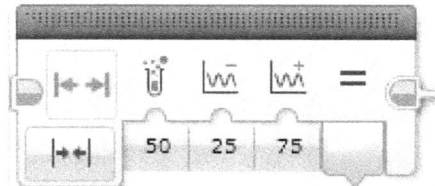

Wire the output of the ultrasonic sensor to the test value of the Range block to track the distance between you and the puppy. If the puppy is twenty centimeters away, give or take a few centimeters, then it should move forward. If it is closer or further away than twenty centimeters, then it should stop.

The output of the Range block is a logic value, true or false. Use this logic value to control a switch:

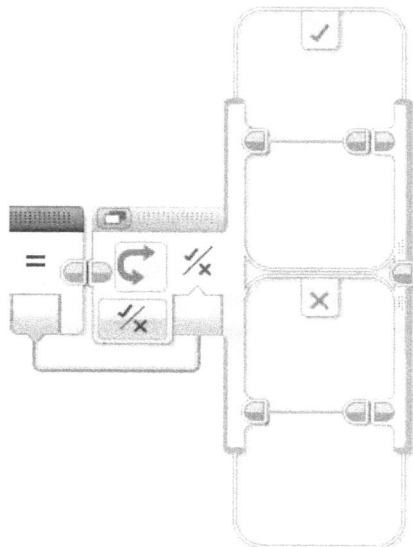

You will need to figure out what blocks to put in the Switch block to make the puppy follow you.

1. Suppose that you switched the Range block from inside mode to outside mode. What other parts of your program would you need to change so that the puppy still followed you at a distance of twenty centimeters?

Teacher Information   Range Puppy

The students use the Range block to program a puppy that will follow them.

Objectives
1. To learn how to use the Range block.
2. To examine different ways of controlling a robot.

Materials
EV3, motors, ultrasonic sensor, LEGO® pieces including wheels, computer, cardboard (optional)

Time: Approximately 30 minutes

Notes
1. This activity is designed to be done as the second in a sequence of activities: Puppy Bot, Range Puppy, Proportional Puppy, and Data Logging Puppy.
2. Since the ultrasonic sensor works best when triggered by a flat surface, you may want to make large cardboard "dog treats" to use in guiding the puppies.

Sample Program for Range Puppy

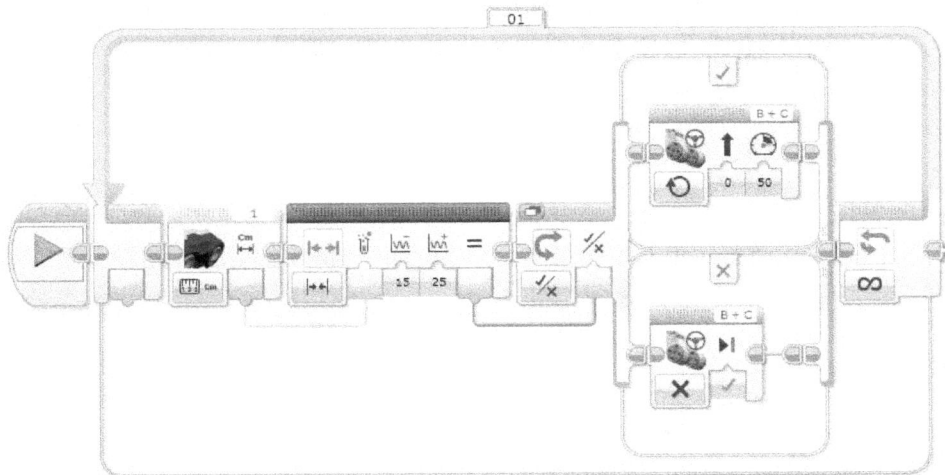

Answers to Range Puppy

1. If the Range block is switched from inside to outside mode, then the contents of the true and false branches of the Switch block must be reversed: motors off if the Switch block is true and motors on if the Switch block is false.

Proportional Puppy

Make a more sophisticated puppy by using a proportional controller to guide its movements. A sample program is shown below.

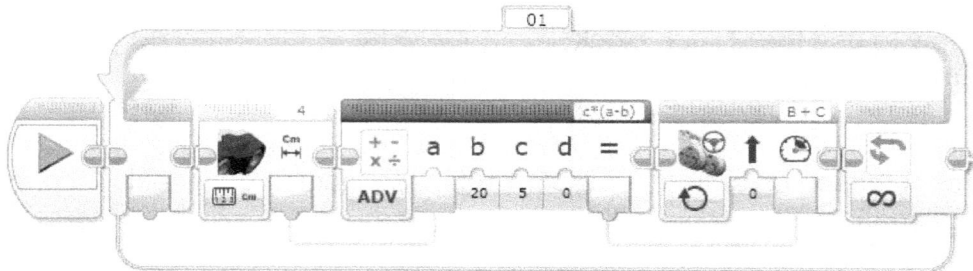

In this more advanced puppy, the motor speed is controlled by an expression instead of by a single sensor input. The difference between the puppy's position as measured by the ultrasonic sensor and the ideal position (in this case 20 cm away) is multiplied by a constant. The result is used as the motor power.

Copy the program above and run it. Try changing b (the ideal distance) and c (the gain) to see how the puppy's behavior changes.

1. Your puppy is scared. How would you change your program to make it follow you more closely? (Try it and see if you are right.)

2. What happens if you make c much larger (greater than 50)?

Teacher Information   Proportional Puppy

The students create a puppy that will follow them.

Objectives
1. To investigate a proportional controller.
2. To see how changes in the feedback gain affect the outcome.

Materials
EV3, motors, ultrasonic sensor, LEGO® pieces including wheels, computer, cardboard (optional)

Time: Approximately 30 minutes

Notes
1. You may want to discuss feedback control with your students before doing this activity. Feedback is often used to control systems. For example, a thermostat uses temperature readings to adjust the level of heat produced: if the room temperature is above the thermostat setting, the heater turns off; if the temperature is above the setting, the heater turns on. The proportional controller in this activity is a little more sophisticated; the distance from the current position to the desired position affects the strength of the response, rather than just turning it on or off.
2. This activity is designed to be done as the third in a sequence of four activities: Puppy Bot, Range Puppy, Proportional Puppy, and Data Logging Puppy (in the data logging section).
3. Since the ultrasonic sensor works best when triggered by a flat surface, you may want to make large cardboard "dog treats" to use in guiding the puppies.
4. If their puppy goes backward, the students will need to adjust their program to multiply the results of their Math block by negative one. An easy way to accomplish this is to reverse the subtraction, using b − a instead of a − b.

Answers to Proportional Puppy

1. Make b (the ideal distance) smaller.
2. If c is made very large, the puppy goes unstable. It will overshoot the ideal position each time, oscillating around it.

Haunted House

Your vehicle must find its way through a haunted house. The house has four rooms; your goal is to visit all four. For each room you reach, you will receive a prize.

The rooms will consist of black rectangles on a white background. The rooms will be numbered one through four; you must visit them in order. To "visit" a room, both of your car's drive wheels must enter the room (though not necessarily simultaneously). Once you have visited all four rooms, you may leave through the front door to collect an additional prize.

Your vehicle must be autonomous; that is, it must drive without any input from you. For example, you could not use a touch sensor to signal the car while it is in the house. However, you may use any sensor(s) you wish to help the car drive itself. If your vehicle drives completely off the white paper or visits a room out of order, your turn is over.

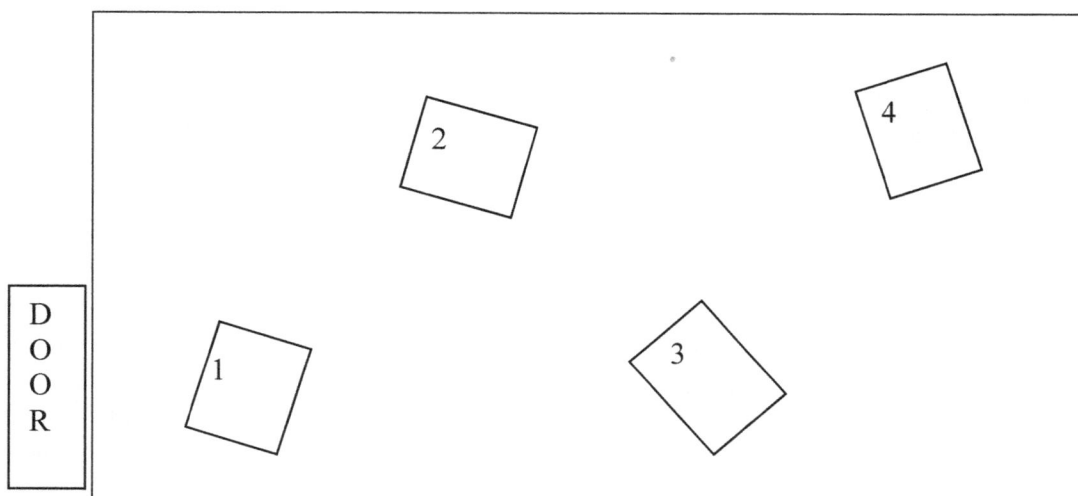

Good luck!

Teacher Information   Haunted House

The students build vehicles and program them to drive through the rooms of a haunted house.

Objectives
1. To program a car to follow a specified route, using rotation sensors, a light sensor, a gyro sensor, timers, or some combination of these.
2. To solve an open-ended problem.
3. To test and modify a program repeatedly.

Materials
EV3 car, light sensor and/or gyro sensor, large sheet of white paper approximately 90 cm by 120 cm, four sheets of black paper approximately 20 cm by 30 cm, prizes, computer

Time: Approximately 50 minutes

Notes
1. To construct the haunted house, start with a large sheet of white paper, approximately 90 cm by 120 cm. (If necessary, you can glue smaller sheets together.) Glue or tape four black sheets on the white field, as shown on the student handout, making sure the edges are securely fastened. Number each room. If you prefer, you can instead use a black house with white rooms.
2. This activity is well suited to the EV3's built-in rotation sensors. The sensors can be used to drive the distance between rooms consistently and precisely.
3. Another possible strategy is to use a light sensor mounted on the car to determine when the car has reached a room. A timer or gyro sensor can be used to accomplish the turns within rooms.
4. Another method that can be used to guide the car is a timer, where the car is programmed to drive for a specified number of seconds, and then turn for some number of seconds, then drive, and so on. This strategy is initially easier than using a sensor, but gives less consistent results (especially if the battery power declines as the car is tested repeatedly).
5. Let the students try their cars in the haunted house as often as they wish. The process of testing and reprogramming is an essential part of this activity.
6. Have token prizes available for successfully completing each room. (If you do this activity near Halloween, many inexpensive Halloween-themed trinkets are available from novelty companies.) One way of handling the prizes is to award each group a numbered card for each room they visit. At the end of the activity, the cards can be redeemed for prizes.

Acknowledgement
This activity is adapted from a similar one by Merredith Portsmore, Center for Engineering Education and Outreach, Tufts University.

Clean Sweep

The task: Build and program a sweeping machine to clear a large area quickly. The vehicle should have two motors, each controlled by a touch sensor attached to a cable.

The challenge: You will have thirty seconds to clear a large square containing 100 LEGO bricks. You must push all of the bricks completely out of the square in that time.

The catch: Your sweeper must be small enough to fit in a closet for storage. It can be no longer or wider than 30 centimeters. Once the thirty-second timer starts, you may remove the sweeper from the closet, position it, and start driving.

Good luck!

Teacher Information   Clean Sweep

This activity makes use of the programming skills that the students have learned.

Objectives
1. To write a program containing multiple loops and switches.
2. To construct a plow or other device for moving objects.

Materials
EV3, motors, touch sensors, LEGO® pieces including wheels and bricks, tape for making the boundaries of the playing field, computer

Time: Approximately 50 minutes

Notes
1. Push-button Fan in the Fan Club section is a good activity to do before this one. In it, the students learn how to control a single motor with a touch sensor. Clean Sweep extends this idea to two motors. A different touch sensor runs each of the motors; by pushing one or both of the touch sensors, the driver can make the car turn or go straight.
2. A reasonable playing field is a square one meter on each side, containing 100 2 x 2 or 2 x 4 LEGO bricks.
3. One possible program (shown below) is to have each motor controlled by a touch sensor. When the sensor is not pressed, the corresponding motor is stopped. Another possibility is to have the motors run forward when the touch sensors are pressed and backwards when they are released.

Sample Program for Clean Sweep

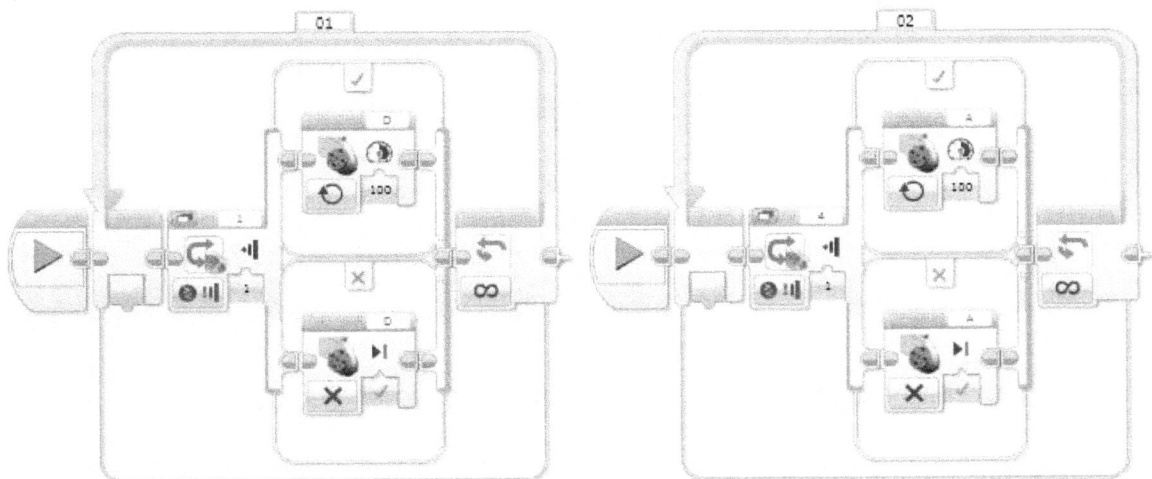

Perfect Pitcher

Build a motorized pitching arm. Can you design an arm that can throw the ball a long distance? Can you design an arm that can throw the ball accurately? An arm that can do both?

The design you use is up to you, but your contraption must contain a lever, a motor, and a touch sensor to start the motor. The motor may be used to power the arm directly or to prepare it. For example, you could use the motor to wind a rubber band, which you would then release to activate the arm.

There will be contests for distance and for accuracy. You may choose to enter either or both of the contests. For each contest, the ball must start behind the tape line. However, the EV3 itself may be in front of the line. You may place the pitching arm however you wish, as long as the part holding the ball is behind the line.

Distance scoring: The distance will be measured from the starting line to the spot where the ball first hits the ground. You will get three tries; the best score of the three will count.

Accuracy scoring: The accuracy target will consist of a cup glued to a plate. Throwing the ball into the inner cup earns three points. Touching the inner cup earns two points. Throwing into the outer plate earns one point. You will get three throws; your accuracy score is the cumulative score for the three throws.

1. List the greatest distance your arm threw, if you entered the distance contest.

2. List your accuracy score, if you entered the accuracy contest.

3. Describe a success you had in building your arm.

4. Describe a difficulty you encountered in building your arm.

5. Make a sketch of your pitching arm. Tell whether it is a first, second, or third class lever. Identify the fulcrum, load (output force), and effort force (input force).

6. Some levers are used to magnify the force. In this case, that would mean that the input force is less than the weight of the ball. Other levers are used to magnify the distance that the load (the ball, in this case) moves. Which type of lever did you build, one that magnifies force or one that magnifies distance?

Teacher Information   Perfect Pitcher

The students design pitching arms for distance and accuracy.

Objectives
1. To design a robotic arm for throwing.
2. To identify types of levers and part of levers.

Materials
EV3, motors, touch sensor, LEGO® pieces, rubber bands, string, plastic ball, plate and cup for target, meter sticks, tape, computer

Time: Approximately 90 minutes

Notes
1. For the distance contest, mark the starting line with tape and position meter sticks along the course so that the distances can be measured easily. You may want to use tape or stickers to mark where the balls land.
2. For the accuracy contest, tape or glue a cup to the center of a plate to make the target. For scoring, award three points if the ball lands in the central cup (even if it then bounces out). Award two points if it touches the inner cup without going inside. Award one point if the ball hits the plate but not the inner cup.
3. As written, this activity has the students identify classes of levers. If you wish to de-emphasize levers, this activity can be done with only the first page of the student handout, omitting the questions about levers.
4. As an introduction, you may want to show the students pictures of catapults (third-class levers) and trebuchets (first-class levers) and discuss how they work.
5. Allow the students to practice in the distance and accuracy areas as much as they wish before the contest begins, adjusting their machines as needed.
6. This activity is a good one for encouraging the students to make methodical changes and note their effects. For example, they can vary the length of time the motors are turned on, the position of the fulcrum, or the length of the throwing arm.

Answers to Perfect Pitcher

1-2. Answers will vary.
3. Answers will vary. One common problem is that the students try to increase the accuracy or throwing distance of the arm by lengthening it, to the point of not having enough torque to lift it.
4. Answers will vary. The trebuchet is a first-class lever, with the fulcrum in the middle. The catapult is a third-class lever with the fulcrum at one end, the effort force close to the fulcrum, and the load at the far end of the lever arm.
5. Whether the students build a first-class lever or a third-class lever, it will almost certainly magnify the distance rather than the force.

Different Drummer

Make a motorized drumming machine. You may use <u>one</u> motor and any LEGO pieces you wish.

The machine must include two drumsticks that alternate playing. That is, one drumstick hits the drum, and then the other drumstick hits the drum, then the first again, and so on.

1. Make a labeled sketch of your mechanism below.

2. Adjust your drumming machine so that one drumstick hits the drum twice as often as the other. Make a sketch of your new mechanism.

Teacher Information   Different Drummer

In this activity, the students build a mechanism to produce alternating motion.

Objectives
1. To create a mechanism that will produce alternating motion using only one motor.

Materials
EV3, motor, LEGO® parts, computer, plates or cups for the drums (optional)

Time: Approximately 30 minutes

Notes
1. A number of methods of producing alternating motion are possible. One is to attach two beams or axles in different offset positions on a single wheel or gear attached to the motor. Another method is to have the motor run two offset cams. A third method is to build a seesaw-like device, with each end of the beam acting as a drumstick.

Answers to Different Drummer

1. Answers will vary. Common solutions include offset levers or levers with cams.
2. Answers will vary. One possibility is to use two offset cams under one lever and just one cam under the other lever. The lever with two cams will hit the drum twice as often as the lever with one cam.

Ramp Up

Design and build a car that can climb as steep a ramp as possible.

To measure how steep a ramp your car climbs, you will use the gyro sensor. Write a program that displays the current gyro sensor reading on the EV3 screen while running both motors at full power.

The gyro sensor can be tricky to use. Make sure that it is level and absolutely still when you turn on the EV3 and download the program. Otherwise, the sensor may not initialize properly. You will also need to make sure that the sensor is mounted on your car in the correct orientation—one that produces a positive angle measure as the car climbs the ramp.

Once your program and sensor are working well, test your car.

1. How steep a ramp can your car climb? Record the angle in degrees.

2. Describe a problem your car encountered. (For example, it veered to the left or the wheels slid instead of rolling.)

3. Describe a change you could make to your car. Explain the physics behind your idea—why you think the change will improve your car's performance.

4. Make the change and test your car again. Did the change help?

5. How steep a ramp can your car climb this time? Record the angle in degrees.

6. Describe a problem your car encountered in this trial.

7. Describe another change you could make to your car. Explain the physics behind your idea.

8. Make the change and test your car again. Did this change help?

9. How steep a ramp can your car climb this time? Record the angle in degrees.

10. Make a sketch of your most successful car.

11. What is the steepest angle that your car was able to climb?

Teacher Information   Ramp Up

This activity gives the students practical experience with a number of concepts in physics, including friction, torque, and center of gravity. It also pushes them to make use of the engineering design process—design, build, test, redesign and retest.

Objectives
1. To build a ramp-climbing car and analyze its performance in terms of physical concepts.
2. To gain additional practice using the gyro sensor.
3. To make use of the engineering design process.

Materials
EV3, gyro sensor, motors, LEGO® pieces including wheels, adjustable ramp, computer

Time: Approximately 60 minutes

Notes
1. To construct the ramp, you will need a sturdy board approximately one meter long. To make the angle adjustable, you can support the top of the board on a bookshelf, moving the top from shelf to shelf to adjust the angle. Another possibility is to support the top of the board with a ring stand and clamps.
2. Encourage the students to follow the instructions on the handout, making only one change at a time and then evaluating it. Many of the students will be tempted to implement several of their ideas at once. Point out to them that it will be much harder to determine the effect of each modification if they have made several changes between trials.
3. If the students reach a dead end in trying to improve the vehicle, help them to analyze the problems they are encountering. If the wheels slip, they may want to try a different type of tires. If the car stalls in one place, they may want to gear it down to increase torque. If the car tips backward, they may want to lower its center of gravity.
4. Torque enters into many aspects of this lab. First, the students can increase the torque of their motors by gearing down the cars. Second, increasing the wheel size decreases the force with which the wheel pushes against the ground, since the torque of the motor is more-or-less constant and the distance to the point where the force is applied increases with the radius of the wheel. Third, the cars tend to veer sideways as the ramp gets steeper. Mounting the forward wheel(s) well in front of the EV3 will tend to counteract this tendency by producing a counter-torque.
5. The handout requires the students to make two design modifications and test them. If time permits, the students may continue to refine their designs.

Sample Program for Ramp Up

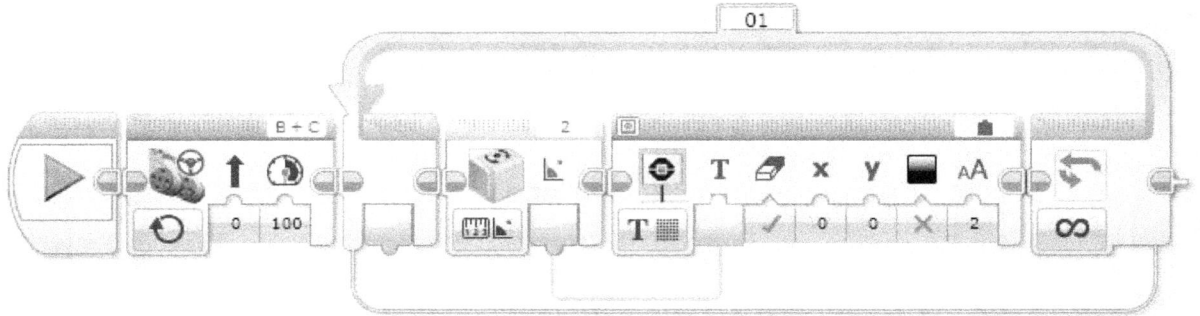

Answers to Ramp Up

Answers will vary. See notes 2 and 3 above for common changes that the students may describe.

Peak Performance

Challenge: You are going on a difficult journey across a valley and up a steep mountain. Build a car to take you there. Your car must climb as steep a mountain as possible. In addition, it must cross the valley quickly.

To measure how steep a ramp your car climbs, you will use the gyro sensor. Write a program that displays the current gyro sensor reading on the EV3 screen while running both motors at full power.

The gyro sensor can be tricky to use. Make sure that it is level and absolutely still when you turn on the EV3 and download the program. Otherwise, the sensor may not initialize properly. You will also need to make sure that the sensor is mounted on your car in the correct orientation—one that produces a positive angle measure as the car climbs the ramp.

Scoring: Your car's score will be calculated by taking the steepest angle your car climbs successfully on the ramp (the mountain) and subtracting the time it takes to cross the floor (the valley). For example, suppose your car crosses the floor in 5 seconds and climbs a 45-degree ramp. Your score would be 45 - 5 = 40.

Good luck!

1. Angle of ramp climbed:

2. Time to cross valley:

3. Final score (angle minus time):

Often, engineers find that two of their goals conflict with one another. For example, making an airplane stronger may increase its weight, making takeoff more difficult. The engineers must make trade-offs, compromising on individual goals to make the overall product work better.

4. Describe a trade-off you faced in this activity. How did you solve the conflict?

Teacher Information   Peak Performance

This activity asks students to grapple with conflicting design goals (speed and ramp-climbing ability) to find a good overall solution.

Objectives
1. To determine the attributes necessary for a good ramp-climbing car.
2. To balance trade-offs when two design goals conflict.

Materials
EV3, gyro sensor, motors, LEGO® pieces including wheels, adjustable ramp, meter stick, tape for marking "valley," computer

Time: Approximately 40 minutes

Notes
1. This activity uses the same ramp as the previous one, Ramp Up. You can do one or the other or both.
2. If the students reach a dead end in trying to improve the vehicle, help them to analyze the problems they are encountering. If the wheels slip, they may want to try a different type of tires. If the car stalls in one place, they may want to gear it down to increase torque. If the car tips backward, they may want to lower its center of gravity.
3. Torque enters into many aspects of this lab. First, the students can increase the torque of their motors by gearing down the cars. Second, increasing the wheel size decreases the force with which the wheel pushes against the ground, since the torque of the motor is more-or-less constant and the distance to the point where the force is applied increases with the radius of the wheel. Third, the cars tend to veer sideways as the ramp gets steeper. Mounting the forward wheel(s) well in front of the EV3 will tend to counteract this tendency by producing a counter-torque.

Sample Program for Peak Performance

Answers to Peak Performance

1-3. Answers will vary.

4. Many students will describe the trade-off they faced between torque and speed. In order to climb a steep ramp, they needed to increase the torque by gearing down. However, gearing down slowed down the car, increasing the time it took to cross the valley. The students needed to decide upon a compromise between these two conflicting goals. Another trade-off that students wrestle with is wheel size. Large wheels make the car go faster. However, they also reduce the force with which the wheels push against the ramp, hurting the car's performance on the ramp.

Applause Meter

Build an applause meter with a piece that spins slowly when the applause is muted, then picks up speed as the applause increases. You may design the spinning part of your meter to be as fancy as you wish.

To program your applause meter, you will need to use the value from the sound sensor to control the motor speed by wiring the sensor value to the power level of the motor.

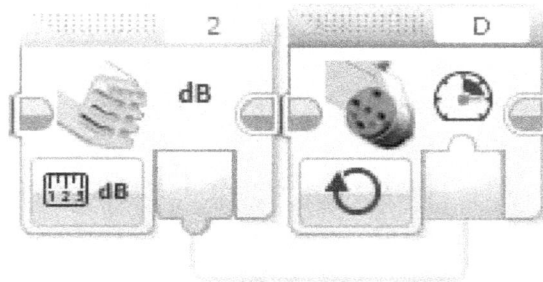

After you have finished your program and built your spinner, test your applause meter. It will probably work best with several people clapping; ask some of your classmates to help you test it.

1. Make a sketch of the spinning part of your applause meter.

Teacher Information   Applause Meter

In this activity, the students construct applause meters that use the noise level to control the motor speed.

Objectives
1. To use the sound level to control the motor power.
2. To gain additional experience in programming and using sensors.

Materials
EV3, sound sensor, motor, LEGO® pieces, computer, craft materials (optional)

Time: Approximately 40 minutes

Notes
1. This activity uses the sound sensor, which is included in the EV3 software, but not in the kit. It can be bought separately.
2. Daytime Fan (in Fan Club) and Puppy Bot (in STEM Activities) both use similar programs.
3. If you wish, you can provide pipe cleaners and other craft materials for constructing the applause meters.
4. By switching the motor block to Move Steering or Move Tank, this program can be used to run a car that responds to the noise level in the room.

Sample Program for Applause Meter

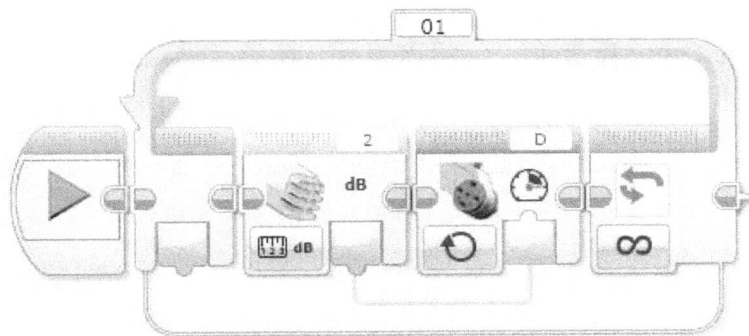

Answer to Applause Meter

1. Answers will vary.

Musical Instrument

Build a musical instrument using either the color sensor or the ultrasonic sensor. Write a program to convert your sensor readings into tones that are audible to people; the human ear can detect frequencies between about 20 and 20,000 hertz. Then download the program and move your EV3 around to create a song.

If you choose to use the color sensor, program the instrument to play higher pitches when the light is bright and lower pitches when the light is dim. If you choose the ultrasonic sensor, program the instrument to play lower pitches when the sensor is close to an object and higher pitches when it is farther away.

For this activity, you will need to use a sensor value to control an output. However, the frequencies detected by the human ear (20-20,000 hertz) are generally much larger numbers than the sensor output. You will need to do some math to adjust the sensor output before wiring it to the sound block.

Download and run your program. Experiment with varying the duration of the note and the conversion factor. See which combination sounds best.

1. What note duration did you choose?

2. What conversion factor did you decide upon?

Extensions

1. Try modifying your instrument and your program to use the other sensor (color or ultrasonic).

2. Try using the color sensor set to detect color rather than light.

Teacher Information   Musical Instrument

In this activity, the students use color or ultrasonic sensors to create music by programming the EV3 to convert sensor readings to musical tones.

Objectives
1. To experiment with changes in pitch and duration.
2. To write a program that uses sensor output to control another process.

Materials
EV3, color sensor, ultrasonic sensor, LEGO® pieces, computer

Time: Approximately 30 minutes

Notes
1. The program required for this activity is fairly sophisticated. Daytime Fan (in Fan Club) is a good activity to do before this one; the program is similar but does not include mathematical manipulation of the sensor values.
2. If you wish to have your students experiment with pitch without writing the program, you can give them the program and have them experiment with changing the duration of the note and the conversion factor from sensor reading to hertz.
3. Have the students run the program tethered to the computer with Port View turned on, so they can see the sensor values change as their instrument plays.
4. Either ambient or reflected light can be used for the color sensor. You might want to have students try both to see how their results are affected.

Sample Program for Musical Instrument, Color Sensor (Light)

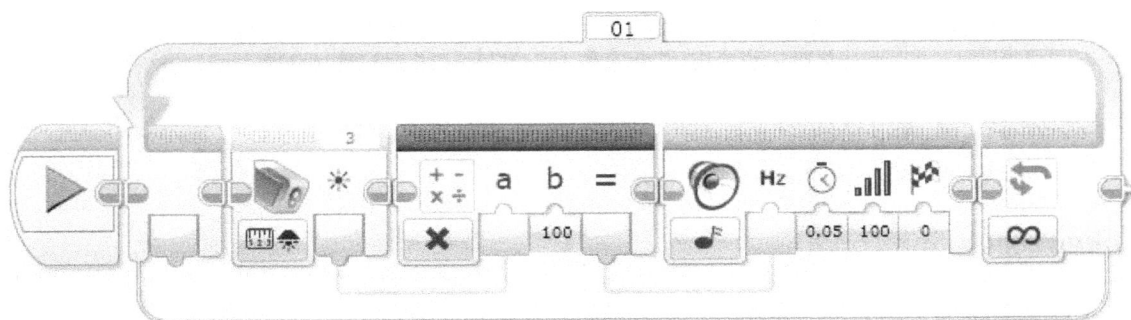

This program multiplies the value of the light reading by 100 to convert it into a reasonable value for hertz. This value in hertz is played for 0.05 seconds (50 milliseconds). The program then repeats.

Sample Program for Musical Instrument, Ultrasonic Sensor

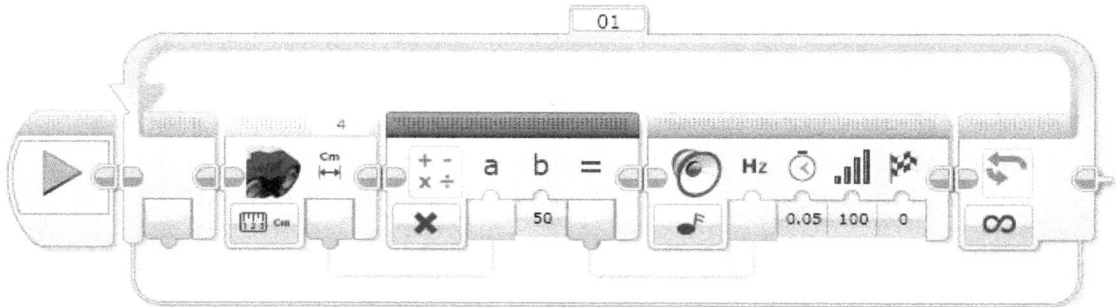

This version of the musical instrument program uses the ultrasonic sensor instead of the color sensor.

Sample Program for Musical Instrument, Color Sensor (Color)

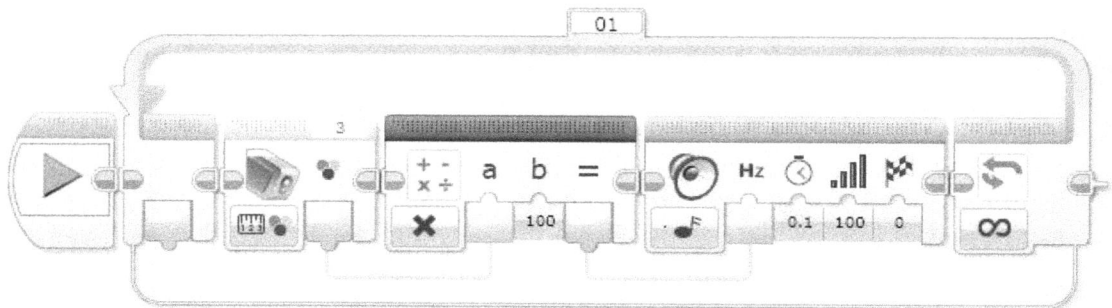

Answers to Musical Instrument

1. Answers will vary.
2. Answers will vary.

Acknowledgement

This activity is based upon one by Professor Chris Rogers of Tufts University.

Random or Not

Some of the EV3s have been programmed to display a random sequence of numbers. In others, the sequence of numbers is not random. Can you figure out which is which?

A random EV3
• is equally likely to display 1, 2, 3, 4, 5, or 6 each time
• does not follow any sort of repeating pattern
• has each number independent of the ones that came before.

Try each EV3 and note your results below. To test an EV3, start the program running and press the center brick button. Each time you press the button, you will hear a tone and see a number on the display screen. Generate a sequence of numbers and try to decide if the EV3 is random or not.

Report the results of each run as a string of numbers. For example, the string 1332… would mean that the first time you pressed the brick button, the EV3 displayed the number one, the second time it showed a three, the third time it a three again, and the fourth time it showed a two. Test the EV3 as many times as necessary to decide if the pattern is random or not. If you start a new run, indicate it with a slash. Good luck!

Label color:

Results:

Random or not?

If not, why not?

Label color:

Results:

Random or not?

If not, why not?

Label color:

Results:

Random or not?

If not, why not?

_____

Label color:

Results:

Random or not?

If not, why not?

_____

Label color:

Results:

Random or not?

If not, why not?

_____

Label color:

Results:

Random or not?

If not, why not?

_____

Label color:

Results:

Random or not?

If not, why not?

_____

Teacher Information   Random or Not

In this activity, the students use the EV3 to investigate randomness.

Objectives
1. To explore the concept of randomness.
2. To investigate how sample size can influence the conclusions drawn.

Materials
Several EV3s, labels of different colors, computer

Time: Approximately 40 minutes

Notes
1. Before starting this activity, program each EV3. Some bricks should contain the random program; the others should each contain one of the non-random programs outlined below. Mark each EV3 with a different color of label.
2. Encourage the students to sample a long enough sequence of numbers that they are able to draw reliable conclusions.

Sample Programs for Random or Not

Random—this program generates a random sequence of the numbers 1 through 6:

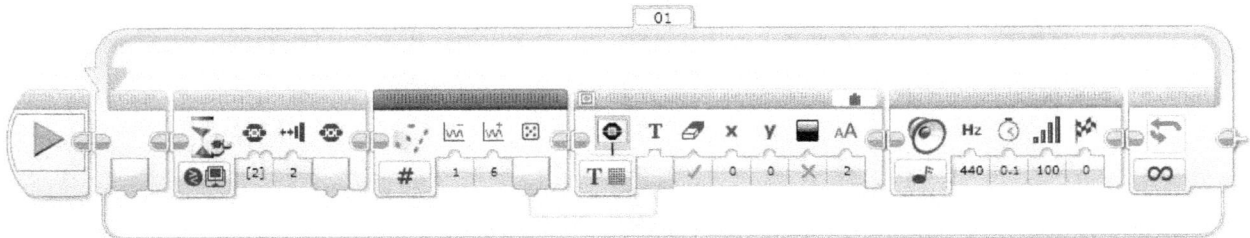

Repeating sequence 1, 2, 3, 4, 5, 6:

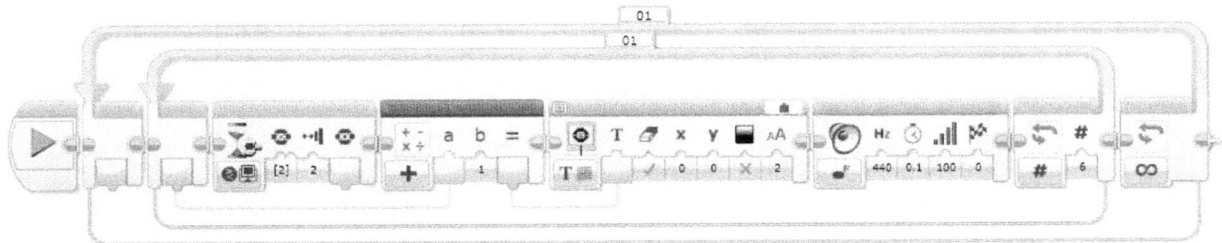

Random sequence of the numbers 1, 2, 3, 4, 5 (no 6):

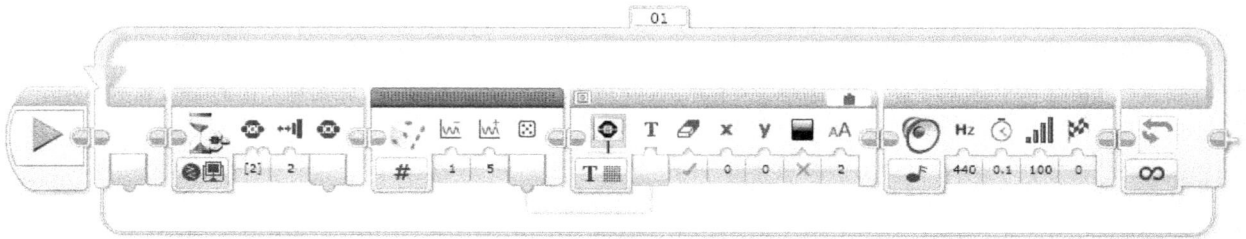

Repeating sequence of a random number followed by the next consecutive number:

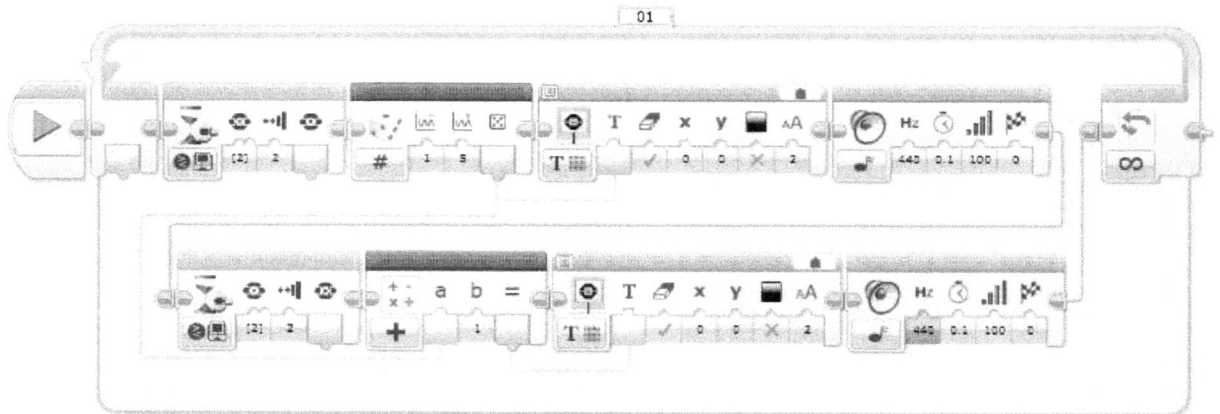

Alternating sequence of a random number and the number 2:

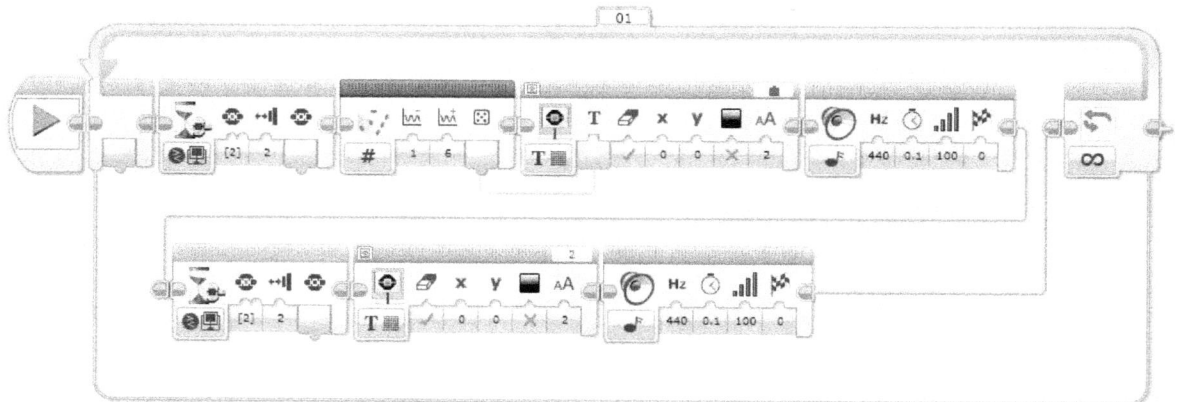

Loaded Dice

The Roll of the Die program shown below generates a random number between one and six, accompanied by a tone, each time the EV3 lower button is pushed.

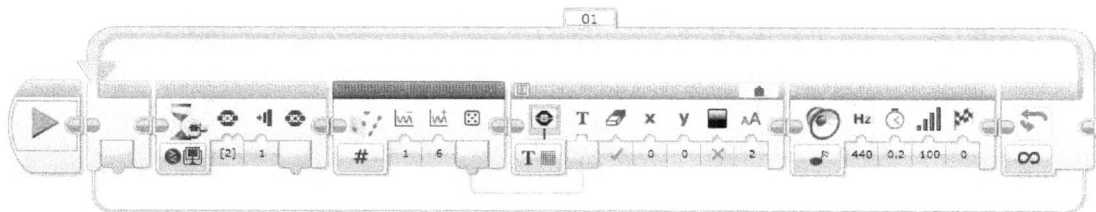

Modify the program to produce a loaded die, one that produces the number six at least half the time, but produces the other five numbers as well, even if rarely. There are lots of ways to accomplish this. List two possible methods below.

1.

2.

Choose one of your ideas and write the program for it. Test it to make sure it works. From the possible algorithms you described above, circle the one you turned into a program.

Now invent your own loaded die. It must produce all six numbers, but they cannot all be equally likely.

3. Describe the pattern of numbers that your die will generate.

4. Make a sketch or print a screenshot of your program.

Teacher Information   Loaded Dice

The students use the Random and Math blocks to create loaded dice.

Objectives
1. To gain additional programming practice.
2. To experiment with the idea of randomness.
3. To design original programs.

Materials
EV3, computer

Time: Approximately 30 minutes

Notes
1. This activity is related to the preceding activity, Random or Not, which has the students examine sequences of numbers to determine whether or not they are random. You may want to do that activity right before or right after this one.
2. If you wish, you can give the students additional restrictions, such as that their program must use the random number generator, or that it must use a case structure.

Answers for Loaded Dice

1-2. Answers will vary. Some possibilities:

The following program alternates a random number between one and six with the number six.

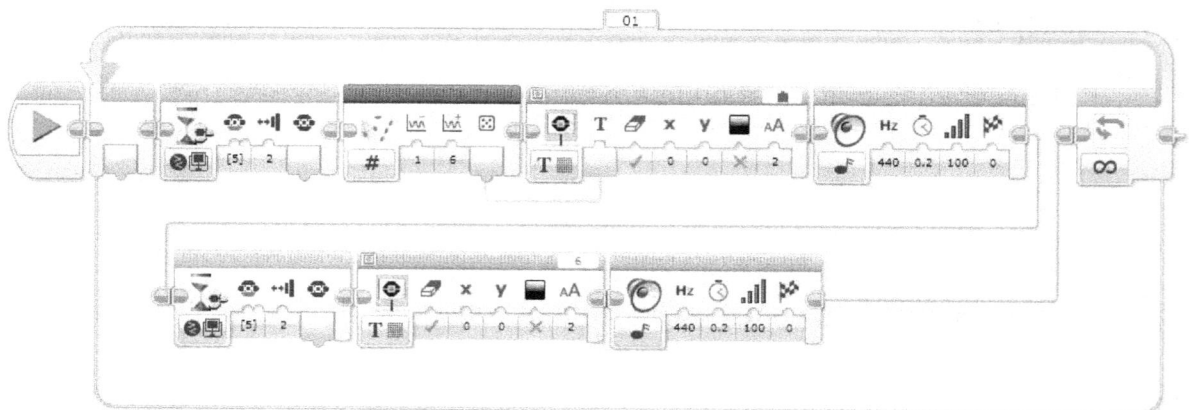

During each iteration, the following program randomly chooses between the true case, which generates a random number between one and five, and the false case, which generates the number six.

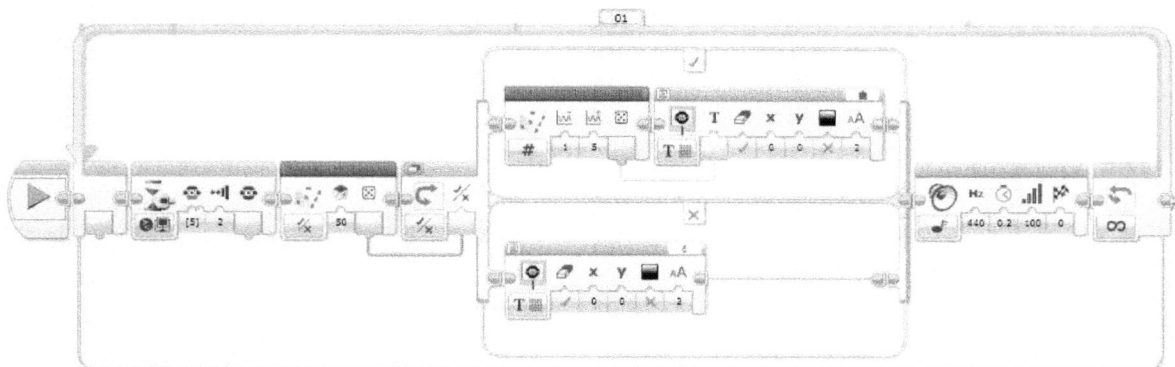

3-4. Answers will vary. See the Teacher's Information for Random or Not for some possible programs.

Efron's Dice

Efron's dice are an unorthodox set of dice with some unusual properties. The four dice are shown below. As you can see, each die is different and none of them has the standard one-through-six numbering.

Die 1:

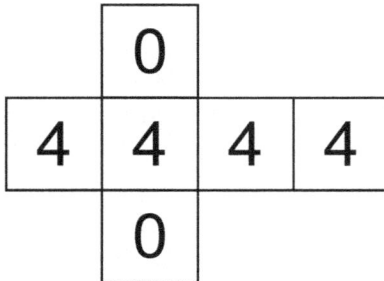

Die 2:

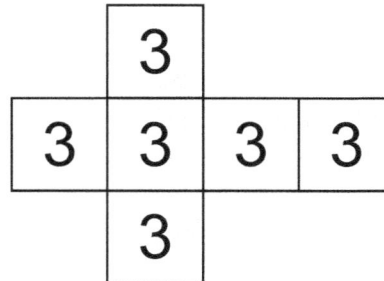

Die 3:

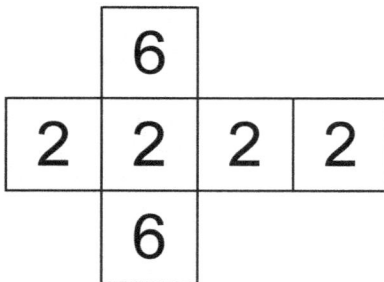

Die 4:

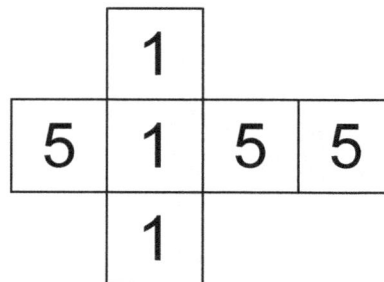

You and a partner are going to use an EV3 version of these dice to play a game. Each person will choose a die by plugging a touch sensor into the corresponding port. For example, to use die 1, plug your touch sensor into port 1 on the EV3. You will each "throw" your die by pressing your touch sensor. When you press the sensor, you will hear a beep and the value of your die will appear on the screen. The person whose die shows a larger number wins a point.

1. Play a short game of five throws each.

Die Used	Points

Now you are ready to play a longer game, twenty throws each. Before you start the game, take a few minutes to decide which die will work best against each other die. For example, let's say you choose Die 1 and your opponent chooses Die 2. If Die 1 rolls a 4 (2/3 chance), then it wins, no matter what Die 2 rolls. If Die 1 rolls a 0 (1/3 chance), then it loses. So, you have a two out of three chance of winning against Die 2. Could you opponent have chosen a better die to play against Die 1?

2. Examine the various combinations and decide which die will work best against each other die. Show your work and explain your reasoning.

Summary:

To beat Die 1, use Die _____

To beat Die 2, use Die _____

To beat Die 3, use Die _____

To beat Die 4, use Die _____

3. Throw twenty times, keeping score as you go. The person with the most points at the end wins.

Die Used	Points

4. Play a second game, switching which player gets to choose first. Record the results.

Die Used	Points

5. Would you rather choose your die first or second in this game? Why?

6. The transitive property says that If A = B and B = C, then A = C. Efron's dice are an example of non-transitive dice. Why do you think they're called non-transitive dice?

Extensions.
1. Find the die with the highest average expected value and the die with the lowest average expected value.
2. Write a program for one of the dice—the program should generate a tone and a number each time the touch sensor is pushed. For example, if you write a program for Die 1, it should randomly generate a 4 two-thirds of the time and a 0 one-third of the time.

Teacher Information   Efron's Dice

In this activity, the students use the EV3 to investigate non-transitive dice.

Objectives
1. To calculate probabilities in order to win a game.
2. To investigate the transitive property.

Materials
EV3, touch sensors, computer

Time: Approximately 40 minutes

Notes
1. Before beginning this activity, download the program to each EV3.
2. Depending upon the mathematical sophistication of your students, you may want to calculate the probable outcomes of one of the dice combinations as a class before having the students calculate the others on their own.

Sample Program for Efron's Dice

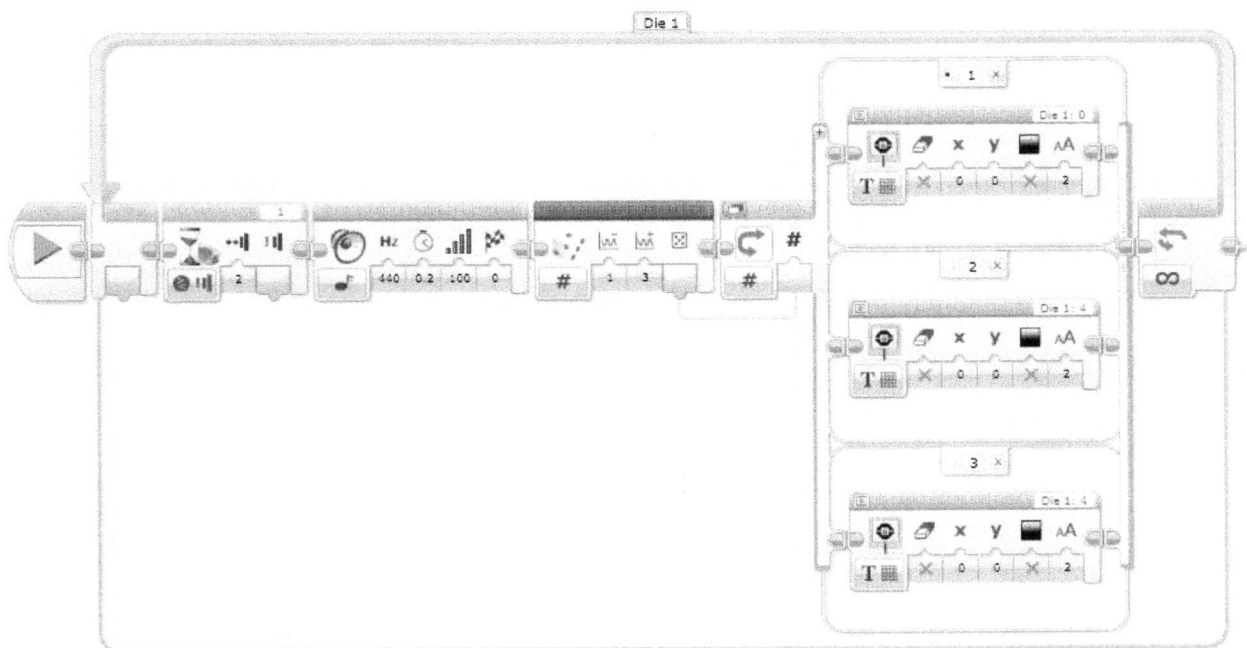

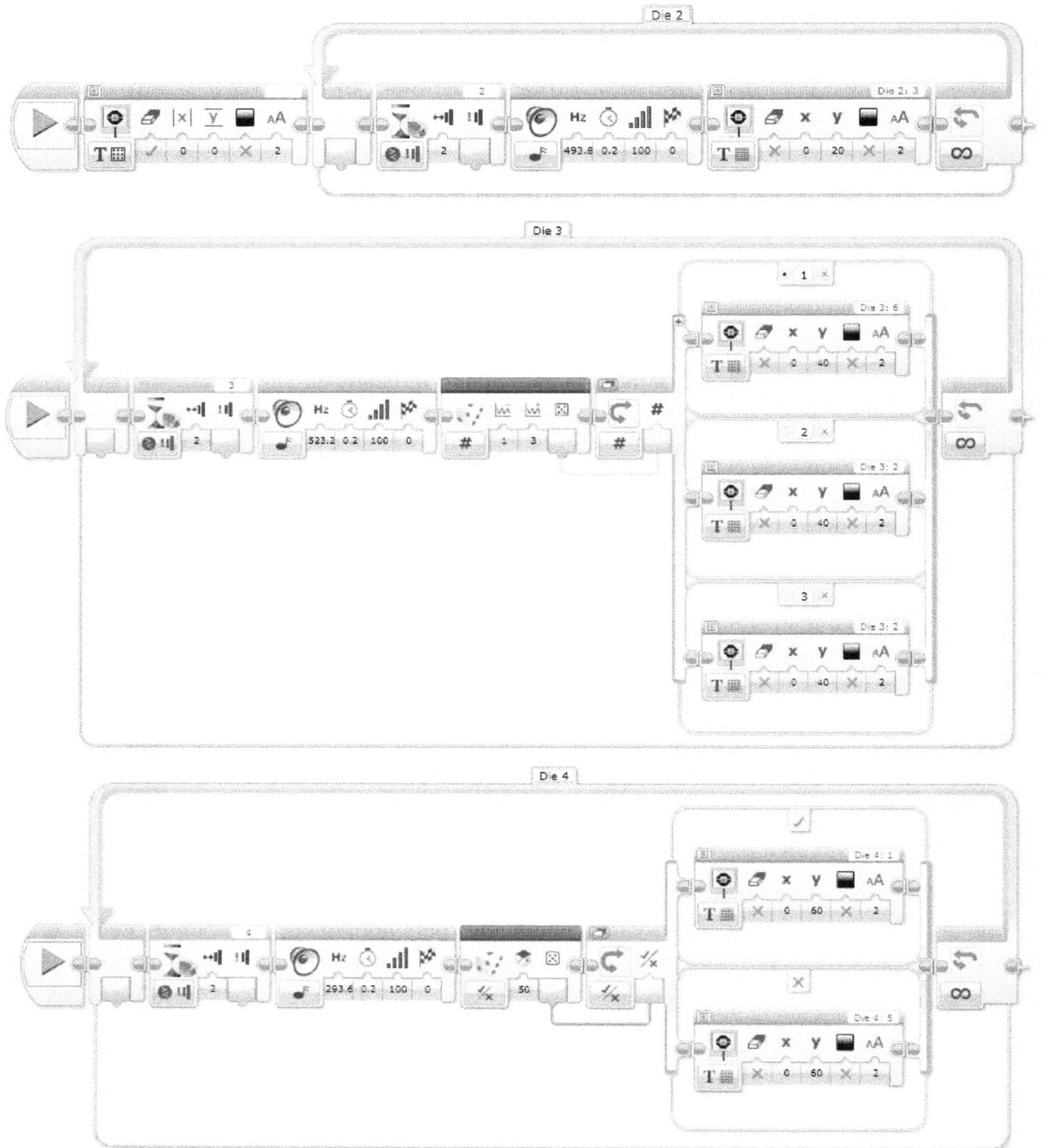

These four loops all run at the same time. The Display block before the Die 2 loop clears the display. Which two loops are shown on the display depends upon which two touch sensors are connected to the EV3.

Answers to Efron's Dice

1. Answers will vary.
2. Summary:

> To beat Die 1, use Die 4
>
> To beat Die 2, use Die 1
>
> To beat Die 3, use Die 2
>
> To beat Die 4, use Die 3

Die 1 versus Die 2: 2/3 chance Die 1 wins, because if Die 1 rolls a 4 (2/3 chance) it wins and if it rolls a 0 (1/3 chance) it loses.

Die 2 versus Die 3: 2/3 chance Die 2 wins, because if Die 3 rolls a 2 (2/3 chance) it loses and if it rolls a 6 (1/3 chance), it wins.

Die 3 versus Die 4: 2/3 chance Die 3 wins. If Die 3 rolls a 6 (1/3 chance), then it wins, no matter what Die 4 rolls. If Die 3 rolls a 2 (2/3 chance) and Die 4 rolls a 1 (1/2 chance), then Die 3 wins. So, Die 3 has a 1/3 chance of winning by rolling a 6 and a 1/3 chance of winning by rolling a 2 (2/3 x ½). Overall, Die 3 has a 2/3 chance of winning.

Die 4 versus Die 1: 2/3 chance Die 4 wins. If Die 4 rolls a 5 (1/2 chance), then it wins no matter what Die 1 rolls. If Die 4 rolls a 1 (1/2 chance) and Die 1 rolls a 0 (1/3 chance), then Die 4 wins. So, Die 4 has a ½ chance of winning by rolling a 5 and a 1/6 chance of winning by rolling a 1 (1/2 x 1/3). Overall, Die 4 has a 2/3 chance of winning.

3-4. Answers will vary.
5. Choosing second is better, because each die can be beaten, on average, by another die. Going second allow you to choose the appropriate die to beat the first player's die.
6. These dice are know as non-transitive dice because even though Die 1 beats Die 2 and Die 2 beats Die 3 and Die 3 beats Die 4, Die 1 does not beat Die 4.

Extensions:
1. Die 3 has the highest average expected value and Die 1 has the lowest expected value. The probability calculations for the dice are shown below.
   Die 1: (1/3 x 0) + (2/3 x 4) = 8/3
   Die 2: 3
   Die 3: (2/3 x 2) + (1/3 x 6) = 10/3
   Die 4: (1/2 x 1) + (1/2 x 5) = 3
   Alternatively, you can find the expected value for each die by adding up the face values and dividing by 6.
2. The programs for the four individual dice are the four loops shown in the sample program. Obviously, Die 2 is the easiest to program, since it always produces the same number.

Voting Machine

A well-designed and accurate voting machine is of vital importance. Build a voting machine using two touch sensors, and then use it to poll your classmates on a question of your choice.

Write a program to keep track of the voting. The program should tally the number of times each of two touch sensors is pressed and display the results on the EV3 screen. To help you get started, here is a program that clears the screen and then displays the number of yes votes registered by a touch sensor in port 2.

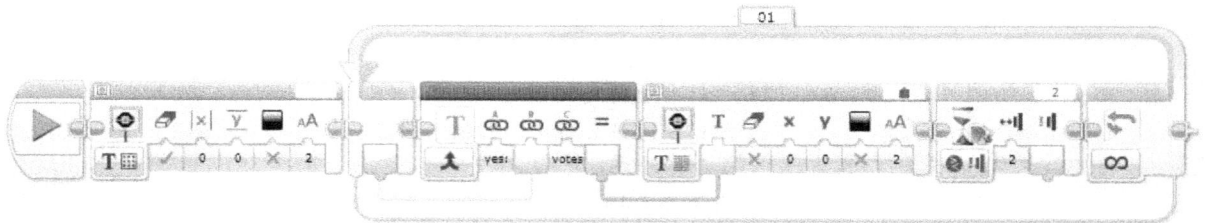

Once your program is working, choose the question you want your classmates to answer. Be sure to choose one with two clear answers, such as a yes/no question or one where the voter chooses between two options. Write the question on a card and label the two touch sensors.

Once your machine is ready, have your classmates vote. Be sure to start the program running before you begin. Before you end the program, make sure that you note the number of votes in each category—the numbers will disappear when you end the program.

1. What question did you put to a vote?

2. Complete the following table:

	Votes for _____	Votes for _____	Total votes
Number of votes			
Percentage of votes			100%

Teacher Information   Voting Machine

In this activity, the students use an EV3 voting machine to collect votes.

Objectives
1. To write a program to tally votes.
2. To consider some of the issues of voting-machine design and use.
3. To calculate percentages.

Materials
EV3, two touch sensors, LEGO® pieces, cards and labels, computer

Time: Approximately 40 minutes

Notes
1. The voting machine can be very simple—the EV3 with two touch sensors attached—or more elaborate, with levers to pull.

Sample Program for Voting Machine

Answers to Voting Machine

1. Answers will vary.
2. Answers will vary. Percentage = (number of votes/total votes) x 100

Do You Have a Sister?

If you ask people if they have a sister, who is more likely to answer yes, males or females? We will run an experiment to find out.

1. First, make a prediction. Do you think more males will answer yes than females, more females than males, or do you think the numbers will be roughly equal?

We will poll a large number of people about whether they have a sister. To collect the data, we will need two voting machines for each group of people we poll, one for males and one for females.

Build two machines with two touch sensors on each. Attach a card containing the question to each machine and clearly label the touch sensors YES and NO. Also be sure to label one machine Males and the other Females to help ensure that the voters use the correct machines.

The program will tally the number of times two touch sensors are pressed, one in port 2 and the other in port 3, and display the results on the screen. Before you end the program, make sure that you note the number of votes in each category—the numbers will disappear when you end the program.

Use your machines to poll a group of people.

2. Once your poll is complete, record your findings in the data table below. Record the data from the other groups as well.

Group polled	# of girls with sisters	# of girls total	% of girls with sisters	# of boys with sisters	# of boys total	% of boys with sisters
Total						

3. What do your data show? Are males or females more likely to have sisters?

4. Do you think this result is true in general? Justify your answer.

Teacher Information

Do You Have a Sister?

In this activity, the students use voting machines to collect data to answer a question about probability.

Objectives
1. To collect large amounts of data in an organized fashion.
2. To solve a probability problem.

Materials
EV3, two touch sensors, LEGO® pieces, cards and labels, computer

Time: Approximately 60 minutes

Notes
1. This activity makes use of the same program as the previous activity, Voting Machine.

Sample Program for Do You Have a Sister?

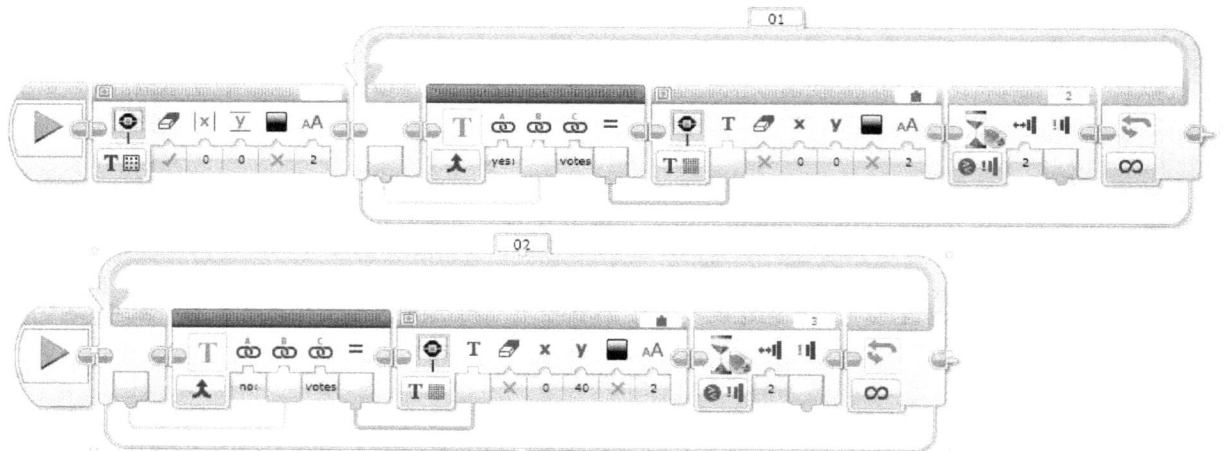

Answers to Do You Have a Sister?

1. Answers will vary. Many students think that males will be more likely to answer yes, because in all families with only one girl, the girl will answer no while the boys will answer yes. (In fact, the numbers on average will be the same, as explained in the answer to question #4 below.)

2. Sample data:

Group polled	# of girls with sisters	# of girls total	% of girls with sisters	# of boys with sisters	# of boys total	% of boys with sisters
Kinder-Garten	6	15	40%	4	13	31%
Third grade	11	16	69%	12	17	71%
Sixth grade	11	27	41%	7	20	35%
Total	28	58	48%	23	50	46%

3. For large data sets, the percentages of males and females with sisters should be about equal.
4. This result holds true in general. To see why, think about all families with exactly two children. There are four equally likely possibilities: two boys, two girls, an older boy and younger girl, and an older girl and younger boy. In the two-boy families, both boys will answer no. In the two-girl families, both girls will answer yes. In the mixed-sex families, the girls will answer no and the boys will answer yes. So, of these eight types of children, half the girls will answer yes and so will half the boys. This line of reasoning holds true for families with more than two children as well.

Reaction Time

How fast can you react to a stimulus? Use the EV3 to measure your reaction time.

Attach a touch sensor to port 1. When the reaction-time program shown below is run, the brick-status light will turn on after a random interval. As soon as it lights up, press the touch sensor as quickly as you can. The program will record how many seconds you took to react and display the result. To end the program, press the center brick button.

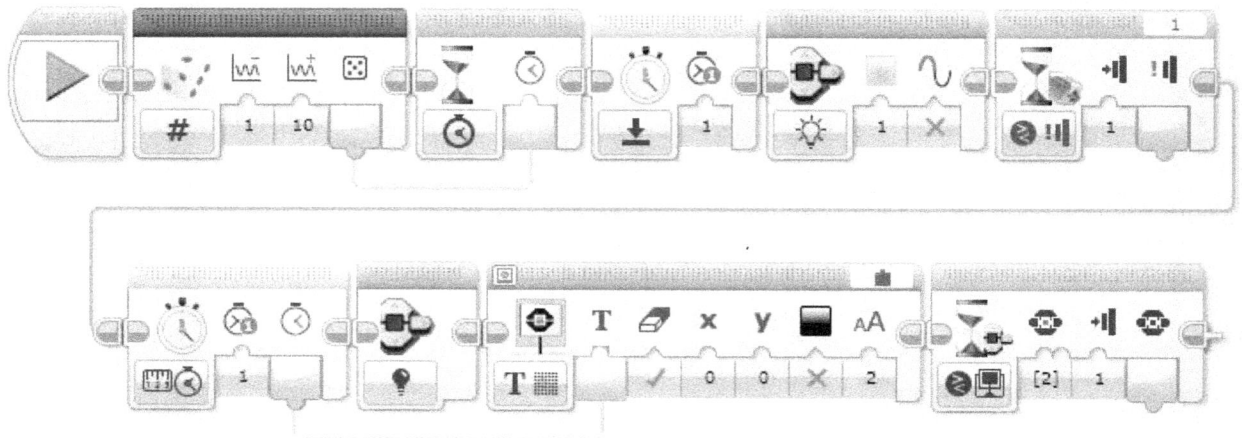

1. Run the program five times, recording your reaction time at the end of each run.

Trial number	Reaction time in seconds

2. What was your fastest time for the light-stimulus program?

3. What was your slowest time for the light-stimulus program?

4. Calculate your average reaction time for the light-stimulus program. What was it?

Do you react as quickly to a stimulus you hear as to one you see? To find out, modify the program to play a sound instead of turning on a light. Be sure to use Play Once instead of Wait for Completion in the Sound block so that the timer starts as soon as the sound begins. Run the program and compare your results to the light-stimulus version.

5. Run the sound-stimulus program five times, recording your reaction time at the end of each run.

Trial number	Reaction time in seconds

6. What was your fastest time for the sound-stimulus program?

7. Your slowest time for the sound-stimulus program?

8. Your average time for the sound-stimulus program?

9. Look at your two sets of data. Does the type of stimulus (light or sound) appear to make a difference in your reaction time?

Teacher Information   Reaction Time

The students investigate reaction time.

Objectives
1. To use the EV3 to measure reaction time.
2. To examine the mean, minimum, and maximum of a data set.
3. To see if the type of stimulus affects reaction time.

Materials
EV3, touch sensor, computer

Time: Approximately 40 minutes

Notes
1. In the handout, the light-stimulus program is given to the students and they modify it for the other half of the activity. If your students are proficient at programming, you can instead have them write both programs from scratch.

Sample Programs for Reaction Time

The light-stimulus program is included in the handout.

Reaction Time Sound-Stimulus Program:

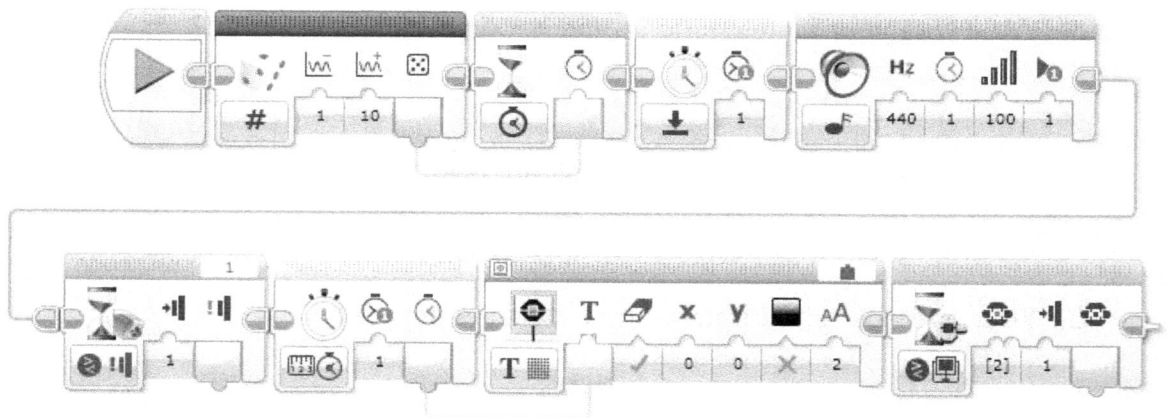

Answers to Reaction Time

1-8. Answers will vary.
9. Answers will vary, though most students should find only small differences between the two stimuli.

Grassfire

Part I: The Theory

A robot cannot plan the same way a person can. Instead, it must have a particular set of rules to follow in order to develop a plan. For example, look at the map below. Suppose that the robot knows where it is and also knows where the goal is--it has a map of some sort. The robot needs a specific set of rules, an algorithm, for planning a route around the obstacle and reaching the goal.

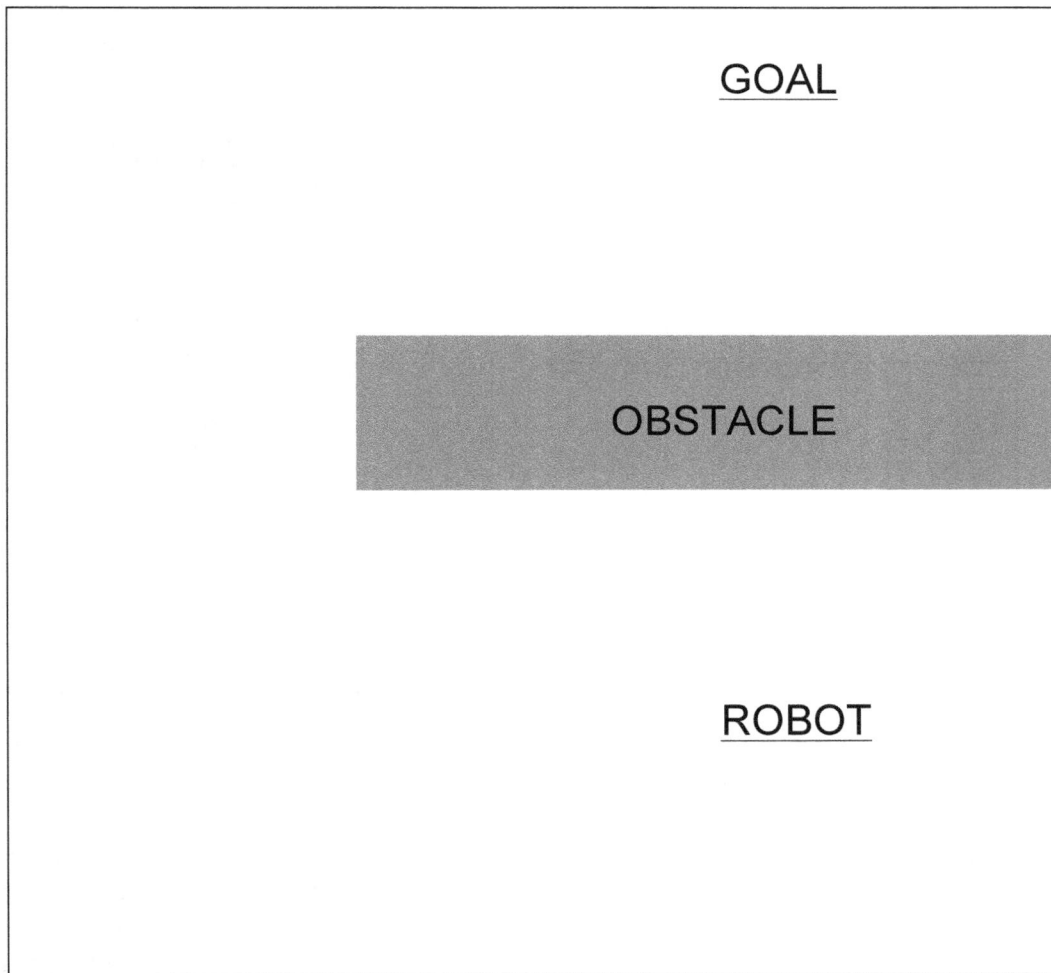

One commonly used algorithm is called the NF1, or grassfire, algorithm. Here's how it works: The map is divided into equal boxes, or cells. The cell containing the goal is given a value of zero. Every cell that is adjacent to the zero cell, either horizontally or vertically (but not diagonally), is given a value of one. Every cell adjacent to a one cell is given a value of two, and so on, until the entire grid is filled. As the grid is being filled, cells containing obstacles are skipped.

1. Use the grassfire algorithm to fill in the grid below.

				GOAL	
		OBSTACLE			
				ROBOT	

Once the grid is filled in, the robot is ready to plan its path. It needs a set of rules for this step also. The rule that the robot uses in the grassfire algorithm is to begin at the robot's starting cell and look for the adjacent cell with the lowest number. For this part of the algorithm, the robot can look at diagonally adjacent cells as well as horizontal and vertical ones. It can also skip numbers. For example, if the robot is in a ten cell and has both a nine cell and an eight cell adjacent to it, it can move directly to the eight cell. The robot moves to the adjacent cell with the lowest number and then repeats the process, moving to successively lower numbers until it reaches zero, the goal.

2. On the map you numbered earlier, use the grassfire algorithm to draw the robot's path to the goal.

3. Was the robot able to reach the goal?

4. How did the path you drew using the algorithm compare to the shortest possible path?

5. Suppose that we changed the algorithm to say that you could not skip numbers as you traced the path. In other words, if the robot were in a ten cell and had both a nine cell and an eight cell adjacent to it, it could not move directly to the eight cell. It would have to move to the nine cell first and then to the eight cell. How would this change affect the robot's path?

Of course, it is possible to vary the size of the cells in the grid. The first grid was six cells across and six cells down. Below is the same map, but this time the grid is only three cells by three cells.

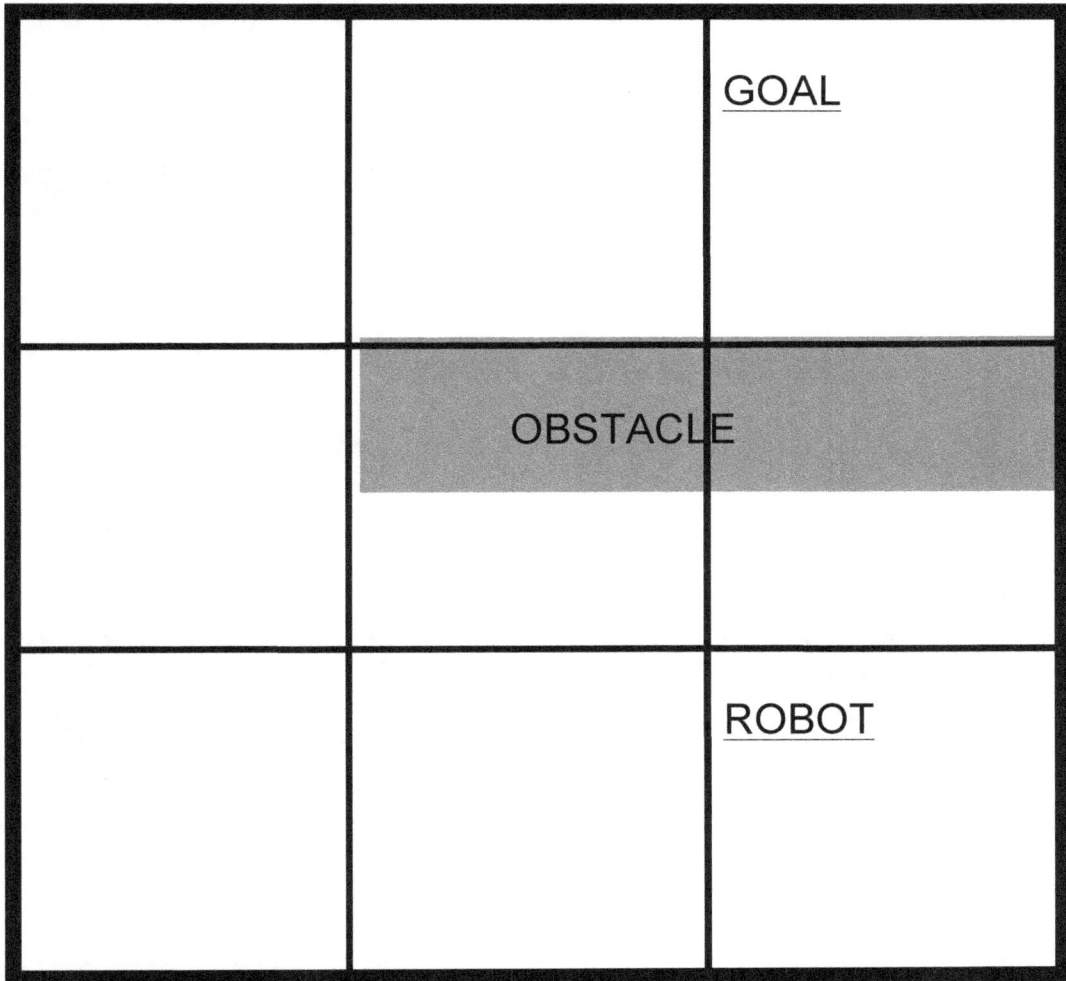

6. Use the grassfire algorithm to number the cells in the grid above and then draw the robot's path. If a cell has a part of an obstacle in it, you may not enter that cell, even if the obstacle does not cover the whole cell.

7. Was the robot able to reach the goal using the three-by-three grid?

8. How did its path compare with the shortest possible path?

9. Name two possible advantages of using a grid with big cells (like the three-by-three grid) over using a grid with smaller cells (like the six-by-six grid).

10. Name two possible disadvantages of using a grid with big cells over using a grid with small cells.

Part II: The Robot

Now that you have seen how the grassfire algorithm works, let's try a simple implementation of it, traversing the three-by-three grid above. To do so, you will write a series of My Blocks: Drive straight between cells, Drive diagonally between cells, Turn left 45 degrees, Turn right 45 degrees, Turn left 90 degrees, and Turn right 90 degrees. Then, you will string together a series of blocks to determine your robot's path.

Here are two simple programs to get you started:

Drive Forward One Cell (30 centimeters):

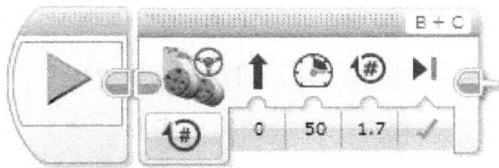

Turn Right 90 Degrees:

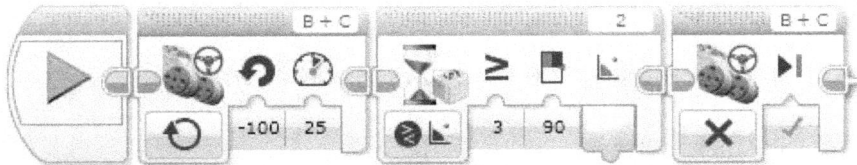

These programs are written for a robot with the motors and gyro sensor oriented forward and with the standard EV3 tires, which have a diameter of 5.5 centimeters. The obstacle course is assumed to have 30-centimeter-long cells. Depending upon your robot and obstacle course, you may need to make adjustments to these programs.

Once these two programs work well for your robot, write the other ones listed above. A couple of hints: The diagonal of a cell is 2.4 times as long as the side. If turning right generates a positive angle on your gyro sensor, then turning left will generate a negative angle.

To make the programs easier to work with, convert each one to a My Block.
Once you are satisfied with your My Blocks for driving and turning, you are ready to build your grassfire program. Add My Blocks to the grassfire program to form the sequence your robot will use to traverse the obstacle course. You will find the My Blocks in the aqua My Blocks palette.

After your program is complete, place your robot in the starting position of the obstacle course and run the program.

11. How did your robot do on the obstacle course? What are some of the shortcomings of this simple building-block program?

Teacher Information   Grassfire

The students investigate the NF1, or grassfire, algorithm for path planning.

Objectives
1. To apply the grassfire algorithm to a map to navigate past an obstacle and reach a goal.
2. To investigate the effects of the fineness of the grid chosen.

Materials
EV3 car, large paper with grid for making the obstacle course, computer

Time: Approximately 80 minutes

Notes
1. Before doing Part II of this activity, you may wish to complete the introduction to My Blocks, Mail Delivery, in Programming Sequences.
2. The programs in the handout are written assuming an obstacle course with 30-centimenter-long cells. Of course, the motor rotations can be adjusted for larger or smaller cells.
3. Depending upon the orientation of the motors and gyro sensor on each car, the motor power and direction and the sign of the angle degrees may need to be adjusted.
4. After completing this introduction, the students can try more complicated maps or try designing their own.
5. Some questions to discuss with the students: If a path to the goal exists, will the grassfire algorithm always find it? (Yes, unless the cell size is large enough that narrow openings are in the same cell as obstacles and thus disappear off the map.) In the case of multiple paths along the grid, will the algorithm always choose the shortest one? (Yes.)
6. The sample grassfire program will allow the students to try the grassfire algorithm, though it is too simple a program to be very accurate, especially for finer grids or more complicated obstacle courses.

Sample Programs for Grassfire

Drive straight across one cell (30 cm):

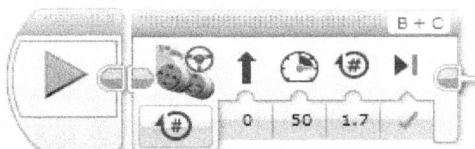

Drive diagonally across one cell:

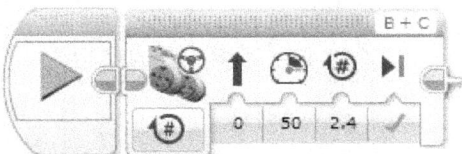

Turn right 45 degrees:

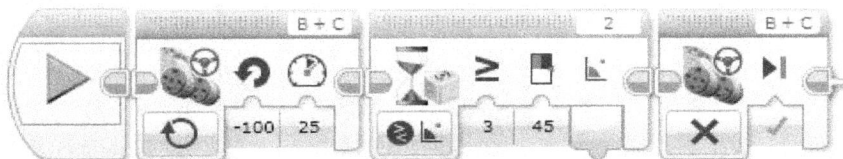

Turn right 90 degrees:

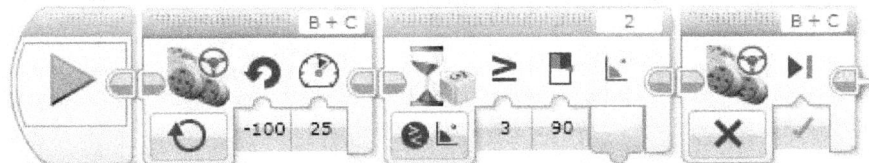

Turn left 45 degrees:

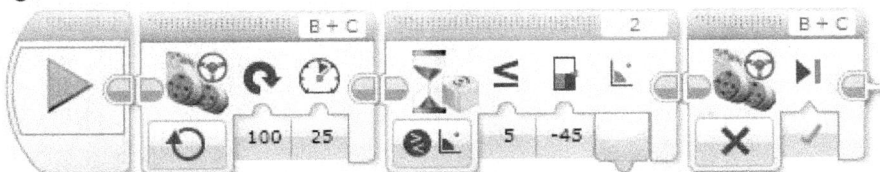

Turn left 90 degrees:

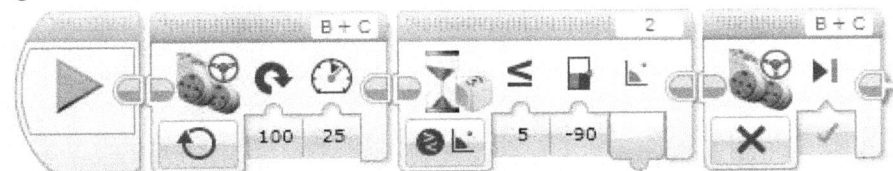

Answers to Grassfire

1. The completed grid:

4	3	2	1	0 GOAL	1
5	4	3	2	1	2
6	5		OBSTACLE		
7	6	7	8	9	10
8	7	8	9	10 ROBOT	11
9	8	9	10	11	12

2. The robot's path using the grassfire algorithm:

4	3	2	1	0 GOAL	1
5	4	3	2	1	2
6	5		OBSTACLE		
7	6	7	8	9	10
8	7	8	9	10 ROBOT	11
9	8	9	10	11	12

3. Yes, the robot was able to reach the goal.

4. The path is close to the shortest possible, except for some added distance to reach the center of each cell.
5. The path would no longer contain diagonal shortcuts, so its total distance would be greater.
6. Three-by-three grid with values shown:

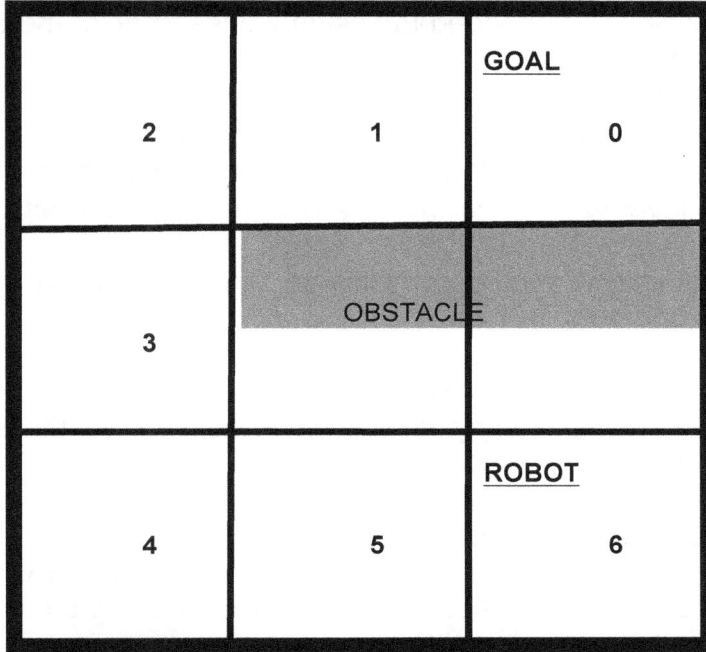

2	1	**GOAL** 0
3	OBSTACLE	
4	5	**ROBOT** 6

Three-by-three grid with robot's path shown:

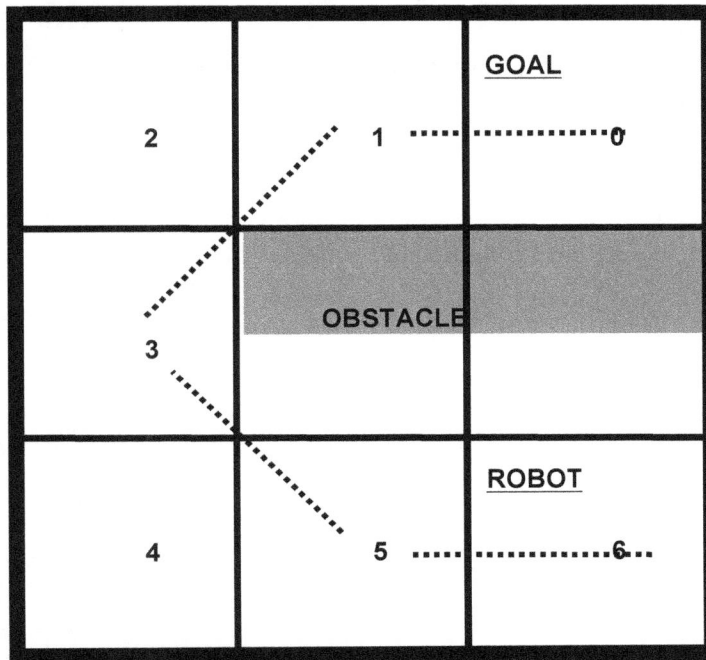

2	1	**GOAL** 0
3	OBSTACLE	
4	5	**ROBOT** 6

7. Yes, the robot was able to reach the goal using the three-by-three grid.

8. Longer, because the new path gave the obstacle a wider berth on one side because of the coarseness of the grid.

9. Possible advantages of using a grid with big cells over using a grid with smaller cells:
   - Fewer steps are needed to assign values to all of the cells and to trace the path.
   - A coarser grid takes less computer memory to store.
   - The larger cells mean that the robot's path is often further from the edges of obstacles, so it is less likely to collide with them.

10. Possible disadvantages of using a grid with big cells over using a grid with small cells:
    - It is difficult to model complicated fields containing multiple obstacles with a coarse grid.
    - The robot may travel further than necessary because it has a wider margin around obstacles.
    - Narrow openings may disappear if they are in the same cell as an obstacle.
    - If the goal is much smaller than the cell, the robot may finish in the correct cell, but still not reach the goal.

11. Answers will vary. A major problem with the building-block program is that the robot does not check the accuracy of its position along the way, so that small errors accumulate.

Blocks Used for Navigating the Grassfire Course

Drive Straight, Turn Right 45, Drive Diagonal, Turn Right 90, Drive Diagonal, Turn Right 45, Drive Straight

Data Logging Activities

These labs introduce the students to the data-logging capability of the Mindstorms software, including Graph Programming. In addition to learning how to use the software to collect data, the students gain experience in analyzing the data they have collected.

The activities in this section are listed below. For each activity, the sensor(s) used and topic(s) covered are listed in parentheses.

The activities in this section are:
1. Light and Dark Scavenger Hunt (color sensor; light)
2. Bragging Rights (color sensor; light)
3. Thunderstorm (color and sound sensors; light, sound, velocity)
4. Crossing the Lines (color sensor; acceleration, distance, velocity)
5. Driving (rotation sensor; circumference, distance, rotation speed, velocity)
6. Deriving (rotation sensor; acceleration, distance, rotation speed, velocity)
7. Zigzag and Diamond (rotation sensor; rotation speed, velocity)
8. Puppy Data Logging (ultrasonic sensor; feedback control, navigation, variables)
9. Which Room? (color sensor; navigation, probability)
10. Stir It Up (temperature sensor; heat)
11. It's a Breeze (temperature sensor; heat)
12. Cool It Fast (temperature sensor; heat)

Light and Dark Scavenger Hunt

Your challenge is to find the highest and lowest light readings you can and record them using the data-logging program.

Start the data-logging program by choosing Add Experiment from the File menu. A data-logging page will open. At the bottom of the page, set the duration of the experiment to 20 seconds and the rate to 10 samples per second. Under Sensor Setup, choose Color Sensor, and then choose Ambient Light Intensity. If you wish, you may also choose your own sensor port and data plot color.

You can run the data-logging program two ways—tethered to the computer, or separate. To run tethered to the computer, use the Download and Run button. As you collect data, it will be displayed on the computer screen. To run separate from the computer, use Deploy. After the program is downloaded to your brick, you can disconnect the USB cable and venture further afield.

To upload your data to the computer, connect your brick to the computer and click the upload button in the lower right corner. You will see all of the available data sets and you can select the one you want to import.

To view the data, click on the Dataset Table button in the lower left corner. You will be able to see all of your data sets, select a color for each plot, and hide or delete data sets.

Teacher Information   Light and Dark Scavenger Hunt

This activity serves as an introduction to data logging. The students learn how to use the Mindstorms software to collect data.

Objectives
1. To learn how to collect and upload data.
2. To interpret the data on a graph.
3. To become familiar with the color sensor.

Materials
EV3, color sensor, computer

Time: Approximately 20 minutes

Notes
1. Doing this lab before any of the following more complicated ones gives the students practice in logging data in a low-stakes lab where data runs can be easily repeated.
2. The students often show considerable ingenuity in registering high and low readings. At the finish of the activity, you may want to give the students a few minutes to share their favorite solutions with the rest of the class.
3. You may want to have the students collect light data using the Reflected Light Intensity to see how it differs from the Ambient Light Intensity setting.

Sample Data for Light and Dark Scavenger Hunt

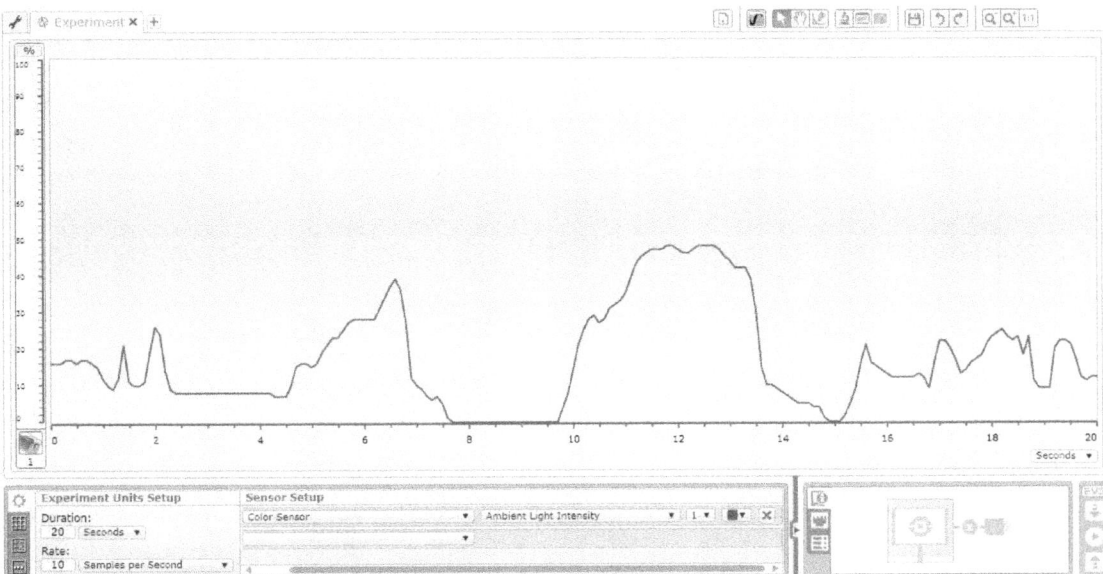

Bragging Rights

The scavenger hunt would be more satisfying if everyone else knew how well you were doing. Modify your data-logging program to trumpet when you have scored particularly high or low light values.

Click the Graph Programming button in the lower left corner. The sensor you have been using will be displayed, along with three zones—star, rectangle, and circle.

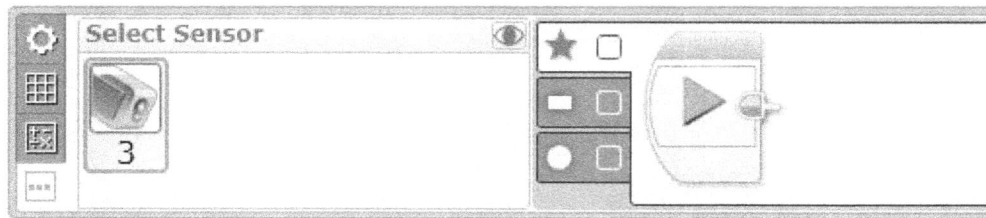

Click on the circle zone and place a check in the box. A line will appear on the graph. You can move this line to determine the boundary of the circle zone. Set it for 10. Whenever your light value goes below 10, the program in the circle zone will run. Click on the programming space at the bottom of the screen to open the programming palette.

Write a program to play a congratulatory sound of some type (your choice) whenever you find a light value below 10. Make your program play a different sound whenever you find a value above 90.

Teacher Information   Bragging Rights

In this activity, the students use Graph Programming to add sound to their light scavenger hunt program.

Objectives
1. To gain additional practice with data logging.
2. To learn how to use Graph Programming.

Materials
EV3, color sensor, computer

Time: Approximately 20 minutes

Notes
1. This activity provides an easy introduction to the Graph Programming function.

Sample Program for Bragging Rights

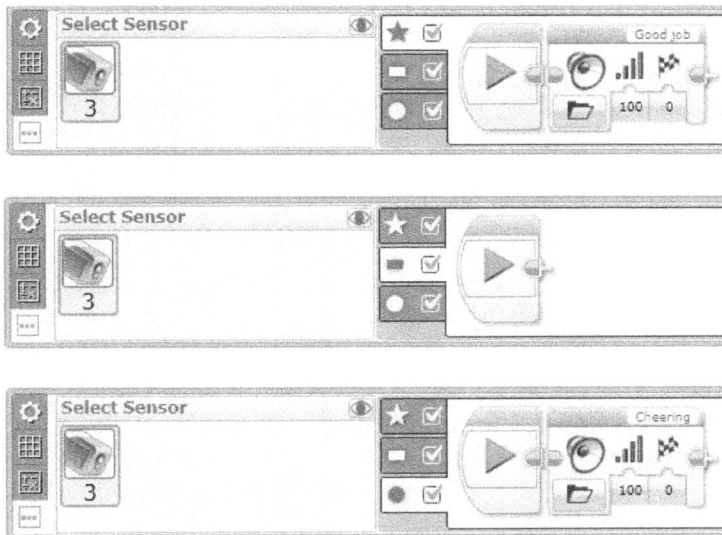

Thunderstorm

If you are near a thunderstorm, you can determine approximately how far away the storm is by counting the number of seconds between when you see the lightning flash and when you hear the thunder.

Here's why:
Light travels very quickly. In fact, the speed of light is 299,792,458 m/s. Even if the lightning flash is several kilometers away, you see it at practically the same instant as it occurs. However, sound travels much more slowly than light. The speed of sound is 340 m/s at sea level. So, if the thunder is several kilometers away, there is a delay between when it happens and when you hear it. You can use the difference between the speed of light and the speed of sound to estimate how far away a thunderstorm is.

1. In one second, how many kilometers can sound travel?

2. If you saw a flash of lightning and three seconds later heard the clap of thunder, approximately how far away would the storm be?

Now you will use your knowledge, along with data logging, to figure out how far away a simulated (fake!) storm is. Attach a color sensor to port 1 of your EV3 and a sound sensor to port 2. Next, write a data-logging program to collect 10 samples per second for 20 seconds. Download the program to your EV3.

When everyone in the class is ready, the teacher will have you start your EV3 and take light and sound readings during a "thunderstorm." After the storm, upload your data to your EV3.

3. How many seconds apart did the lightning and thunder occur?

4. Approximately how far away was the storm?

Teacher Information   Thunderstorm

In this activity, the students log two different sensors at the same time.

Objectives
1. To gain additional practice in using data logging.
2. To learn how to log two different sensors at the same time.
3. To be able to analyze a graph containing multiple lines.

Materials
EV3, color sensor, NXT sound sensor, computer

Time: Approximately 20 minutes

Notes
1. This lab makes use of the NXT sound sensor, which may be purchased separately.
2. To simulate lightning, you can turn the classroom lights on and off. For thunder, any loud, sudden sound—a dropped book, for example—will work.
3. The data for this lab can be collected either of two ways. The first method is to leave the EV3 connected to the computer and have the data displayed in real time. The second way is to disconnect the EV3, collect the data, and then upload them afterwards.

Sample Program for Thunderstorm

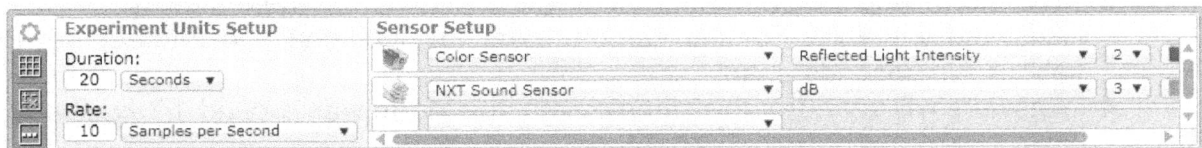

Experiment Units Setup	Sensor Setup			
Duration: 20 Seconds ▼	Color Sensor ▼	Reflected Light Intensity ▼	2 ▼	
	NXT Sound Sensor ▼	dB ▼	3 ▼	
Rate: 10 Samples per Second ▼	▼			

Answers to Thunderstorm

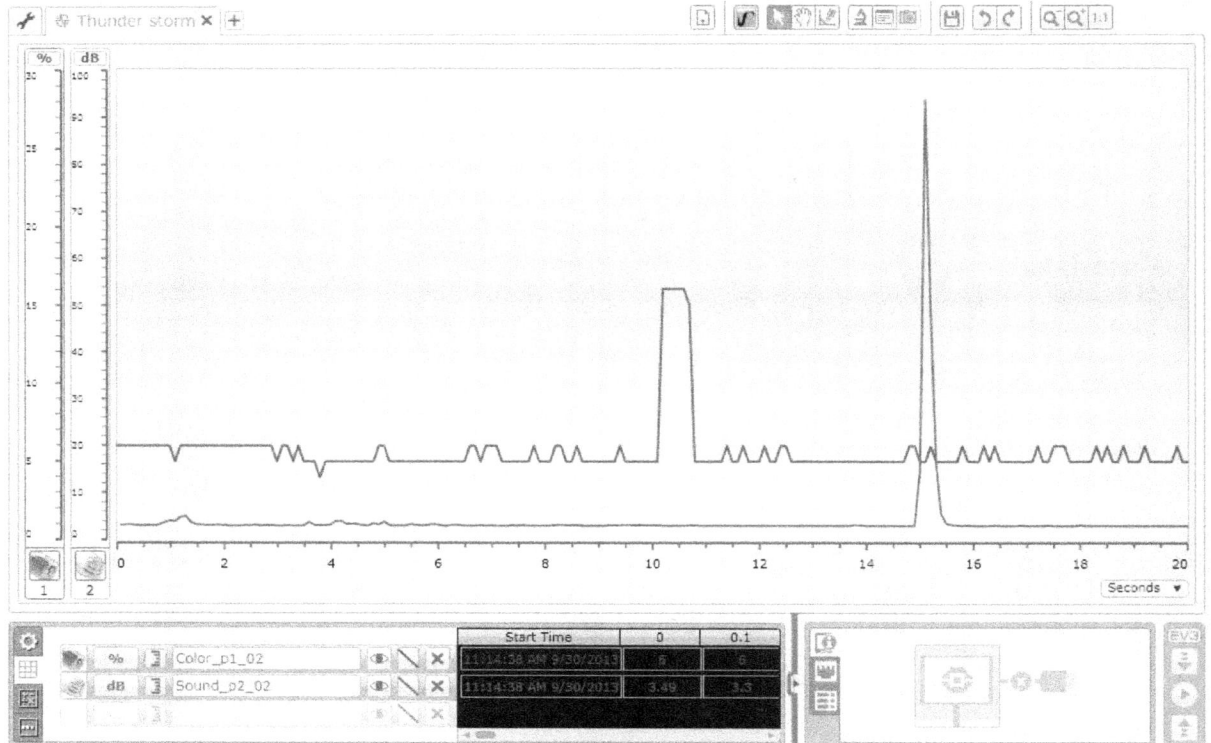

Sample data for a simulated storm.

1. In one second, sound can travel 340 m = 0.34 kilometers, or approximately one-third of a kilometer.
2. The storm would be around 1.0 km (= 3 s X .34 km/s) meters away.
3. Answers will vary. For the sample data, the lightning and thunder occurred about five seconds apart.
4. Answers will vary. For the sample data, the storm was around 1.7 km away.

Crossing the Lines

Use a color sensor to investigate the motion of your car.

Build an EV3 vehicle and attach a color sensor to it. The color sensor should be on the front of the vehicle, close to the floor and pointing downward.

Write a program to log the color sensor as your car moves. Take twenty readings per second. Turn on motors A and C at a power level low enough so that your vehicle does not cross the tape lines too quickly. Run the motors long enough that the car crosses the final line before stopping.

Place your car with the color sensor just behind the starting line. Start the program running and let the vehicle take data.

Return to your computer and upload the data you have collected.

1. How can you tell when your car crossed each line?

2. Does your car accelerate? How do you know?

3. Calculate your vehicle's average speed over the 2.00-meter track. Be sure to show your work.

4. What possible sources of error exist in this experiment? Make a list.

Teacher Information   Crossing the Lines

In this activity, the students use color sensors to measure the speed of their cars, by running the cars over a striped course and analyzing the minima created in the sensor data as the cars cross the dark lines.

Objectives
1. To use a color sensor and data logging to analyze speed and acceleration.
2. To calculate average speed from time and distance.
3. To interpret a graph to see when the car was accelerating or decelerating.

Materials
EV3, color sensor, motors, LEGO® parts including wheels, tape and meter stick for constructing the course, computer

Time: Approximately 40 minutes

Notes
1. Before starting this activity, create a two-meter-long course for the cars. The course should consist of a light-colored background, with a strip of dark tape placed every 20 cm along it. (You can also use light tape on a dark background; the resulting light-intensity measurements will show a maximum for each tape line rather than a minimum.)
2. In order to make the graph easier to decipher, you may want to mark the beginning and end of the track with sheets of dark paper rather than lines.
3. Some students may need help in analyzing the graph. In particular, they may need help in making the connection between wider spacing of the minima and slower speed.

Sample Program for Crossing the Lines

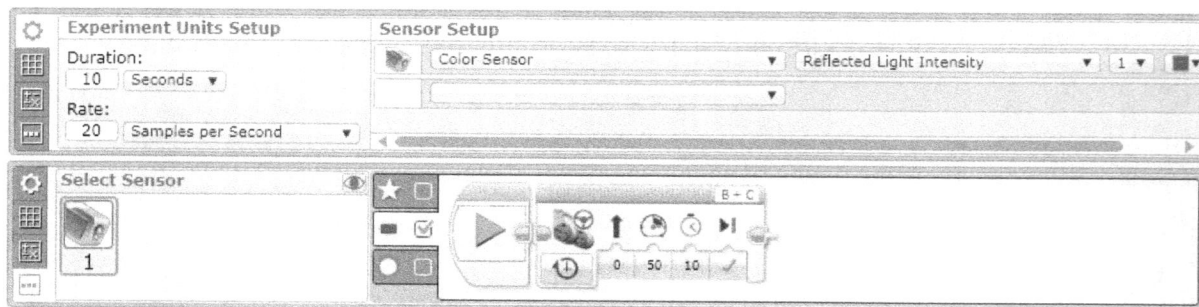

Answers to Crossing the Lines

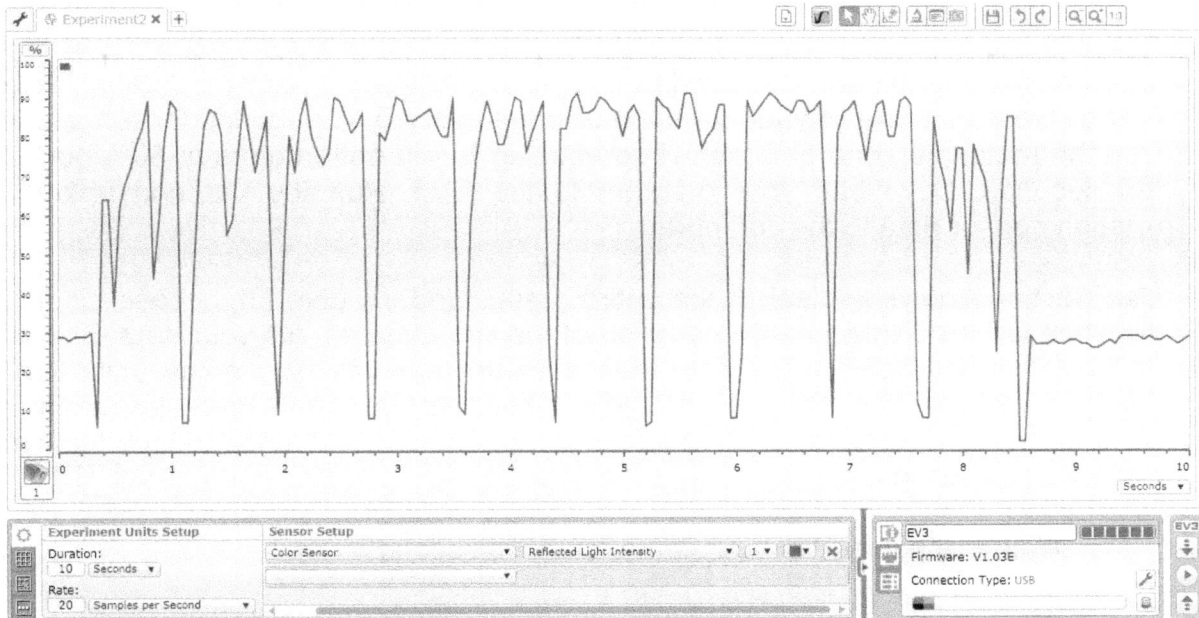

Sample data: Eleven minima are visible, corresponding to the eleven dark lines on the track.

1. The color-sensor graph shows a minimum (valley) whenever a line is crossed (or a peak, if light tape is used).
2. Answers will vary. After a brief initial period of acceleration, most cars will show a constant speed. If the valleys are evenly spaced, then the car is traveling at a constant speed. If the valleys get closer together, then the car is speeding up. If the space between valleys grows wider, then the car is slowing down.
3. Answers will vary. To calculate the average speed for a run, divide 2.00 meters by the time elapsed between when the car crosses the starting line and when it crosses the finish line. For the sample data shown in the graph, the average speed is 0.25 m/s.
4. Possible sources of error include: the track is not exactly 2.00 meters long, the tape lines are not evenly spaced, the car does not travel in a straight line, the color sensor misses a tape line between logging points, the times are not read off the graph correctly, the calculations are not performed correctly.

Driving

What is the velocity of your car? Log the rotation sensor to find out.

Open a new experiment. Set the data logger to take rotation-sensor data in rotations (not degrees) on one of your car's motor ports every 0.05 seconds for ten seconds. Use the rectangle zone of Graph Programming to run both motors at 50% power for ten seconds. Download your program to the EV3. Run the car and upload the rotation-sensor data to the computer.

Use Section Analysis to select the entire graph. Under Curve Fit, choose Linear. An equation will be displayed, the equation of the line that best fits your data. The slope of the line (the coefficient of x in the equation) tells you how many units the line changes vertically for every unit it changes horizontally. In this case, it is telling you the number of rotations every second—in other words, the speed of the car in rotations per second.

1. What is the speed of your car in rotations/second?

2. What do you think will happen to the slope of the line if you increase the motor power to 100%?

3. Increase the motor power to 100% and run the car again. What is the new slope—that is, the speed in rotations/second?

4. When you increased the power from 50% to 100%, did your car go twice as fast? How can you tell?

Now, calculate the velocity in centimeters per second for the car running at full power. To do this, you need to know how far the car travels with each complete rotation of the wheels. Use a marker to draw a heavy ink line across the tread of the tire. (If your car has more than one size of wheels, make sure you use the wheel attached to the rotation sensor.) Roll the car across a piece of paper. The inked wheel should leave a mark each time the ink line rolls over the paper. Measure the distance between two successive lines in centimeters to find the distance per rotation. (Notice that you've just found the circumference of the tire.)

Multiply your full-power velocity in rotations/second by the tire circumference to obtain the velocity in centimeters/second.

5. What is the circumference of your tire?

6. What is the velocity of your car in centimeters/second, running at full power?

Teacher Information   Driving

In this activity, the students use rotation sensors to analyze the speed of their cars, making use of the Analysis Tools to fit a line to their data.

Objectives
1. To use a rotation sensor and data logging to analyze velocity.
2. To use Curve Fit to fit a line to a set of data.
3. To interpret a slope in an equation and on a graph.

Materials
EV3 car, marker, computer

Time: Approximately 30 minutes

Notes
1. If the students have EV3 cars from the previous activity, they can use them for this activity as well.
2. Depending upon the students' background in mathematics, you may want to introduce or review equations of lines and slope with them at the start of this activity.
3. If the car produces downward sloping lines because of the way the motors are oriented, it can be run backwards for this activity.

Sample Program for Driving

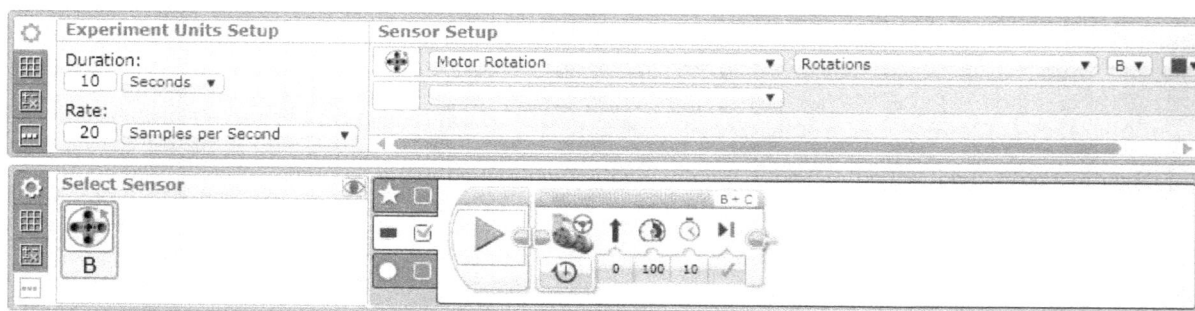

Data-logging program for motors running at full power.

Answers to Driving

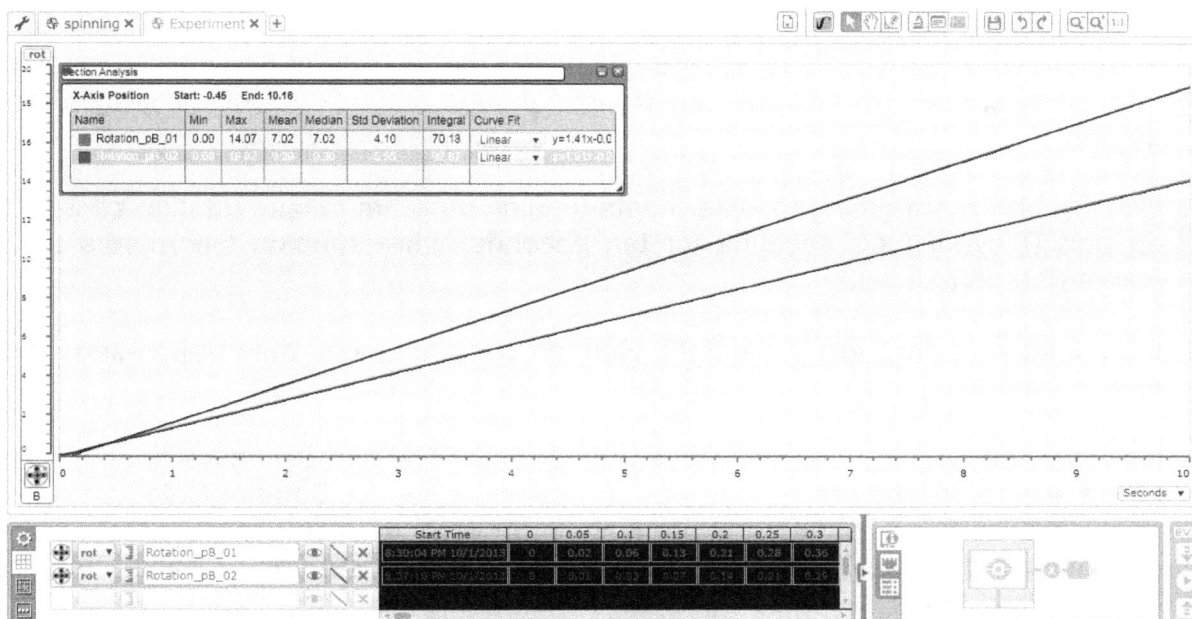

Sample data, showing lines for full power and half power.

1. Answers will vary. In the sample data above, the average speed at half power is 1.41 rotations/second.
2. Answers will vary. Most students will say that the slope of the line will increase. Many will speculate that the slope will double. (Accept any ideas, since they are speculating.)
3. Generally, the slope of the line will increase. In the sample data above, the average speed at full power is 1.91 rotations/second.
4. The speed did not double, since the slope of the line is not twice as steep.
5. Answers will vary, depending upon the tire used.
6. Answers will vary. For the sample data shown above, the average velocity was 32 cm/s, based upon a tire circumference of 17 cm and a slope of 1.91 rotations/second in the equation.

Deriving

In the previous activity, Driving, you calculated the velocity of your car by logging its rotation over time. We can also use rotation data to produce velocity and acceleration graphs.

Use Graph Programming to write a data-logging program to take rotation-sensor data on port B every 0.05 seconds for ten seconds, while running the motors at 50% power (the default value).

Download your program to the EV3. Run the car and upload the rotation-sensor data to the computer.

Now, we will take the first derivative of your data. The first derivative finds the slope of the line (and thus the velocity) at each point. Click on Dataset Calculation in the lower left corner. To find the derivative of your rotation-sensor data, click on derivative in the list of functions, then click on the data set you uploaded, then click the Calculate button.

1. Was the velocity constant or did it vary? How can you tell?

2. What was the average velocity of your car?

Now repeat the steps above in order to find the second derivative, but this time, use the first derivative data as your input data. The second derivative tells you the acceleration of the car, in other words, how fast the velocity was changing.

3. Was the acceleration constant or did it vary?

4. What was the average acceleration of your car?

Teacher Information   Deriving

In this activity, the students use rotation-sensor data and the Analysis tools to produce velocity and acceleration graphs.

Objectives
1. To use data-logging analysis tools to produce velocity and acceleration graphs.

Materials
EV3 car, computer

Time: Approximately 30 minutes

Notes
1. Depending upon the students' background in mathematics, you may want to discuss derivatives and their relationship to slope.
2. The rotation data for this lab can be collected in either degrees or rotations.

Sample Program for Deriving

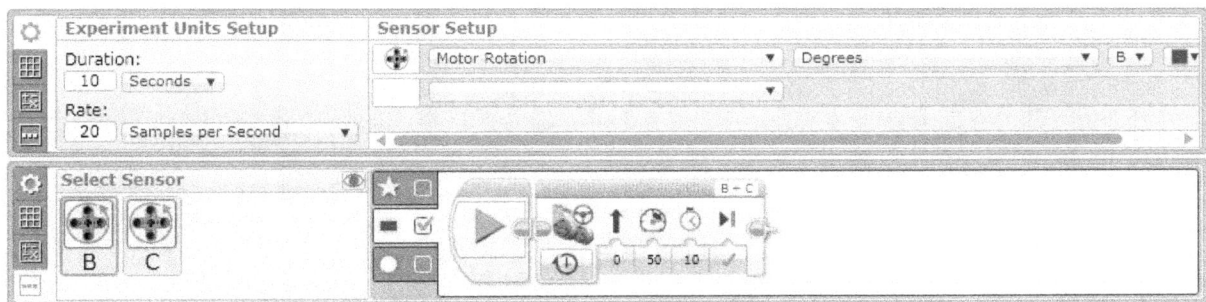

Answers to Deriving

Sample raw data:

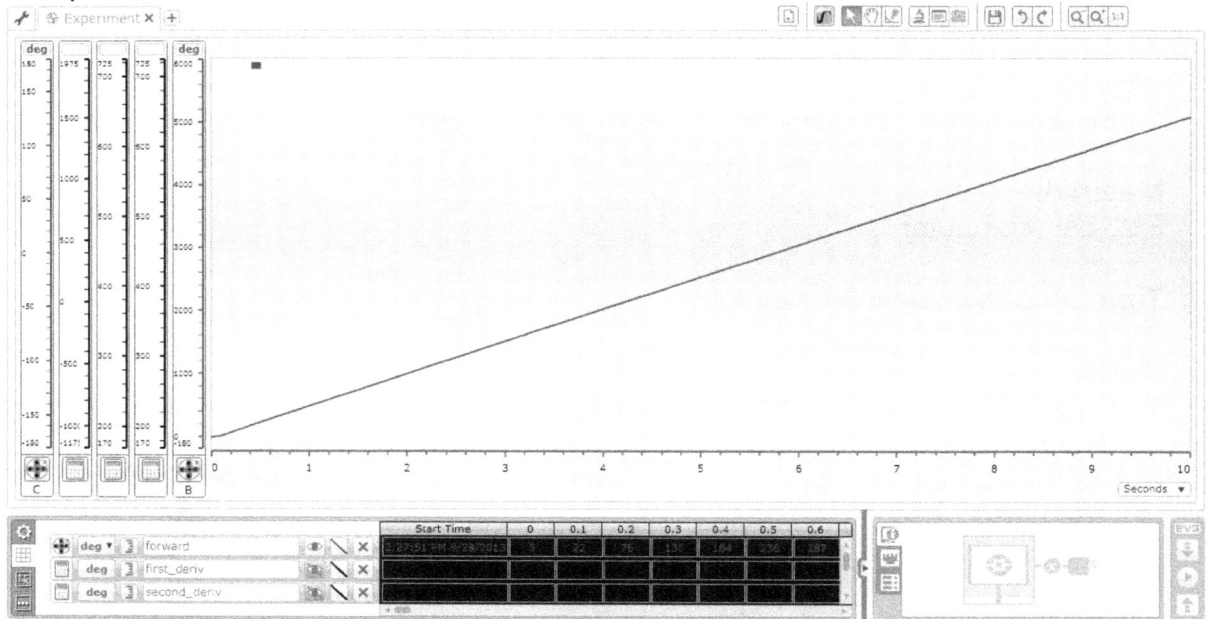

Velocity data (first derivative):

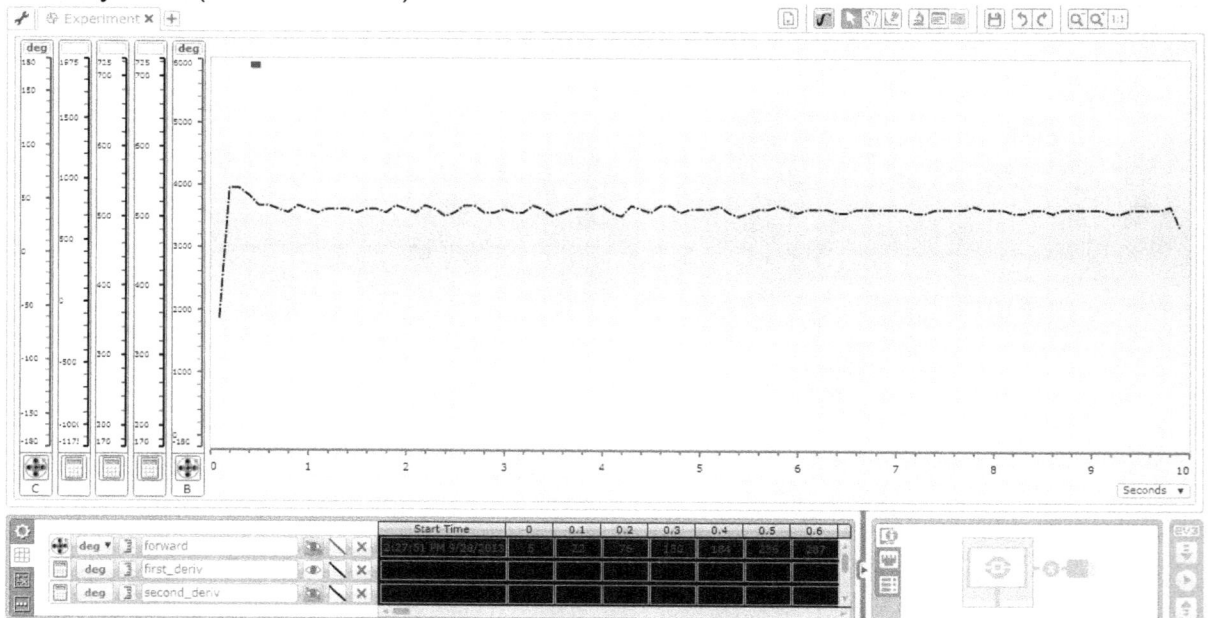

Acceleration data (second derivative):

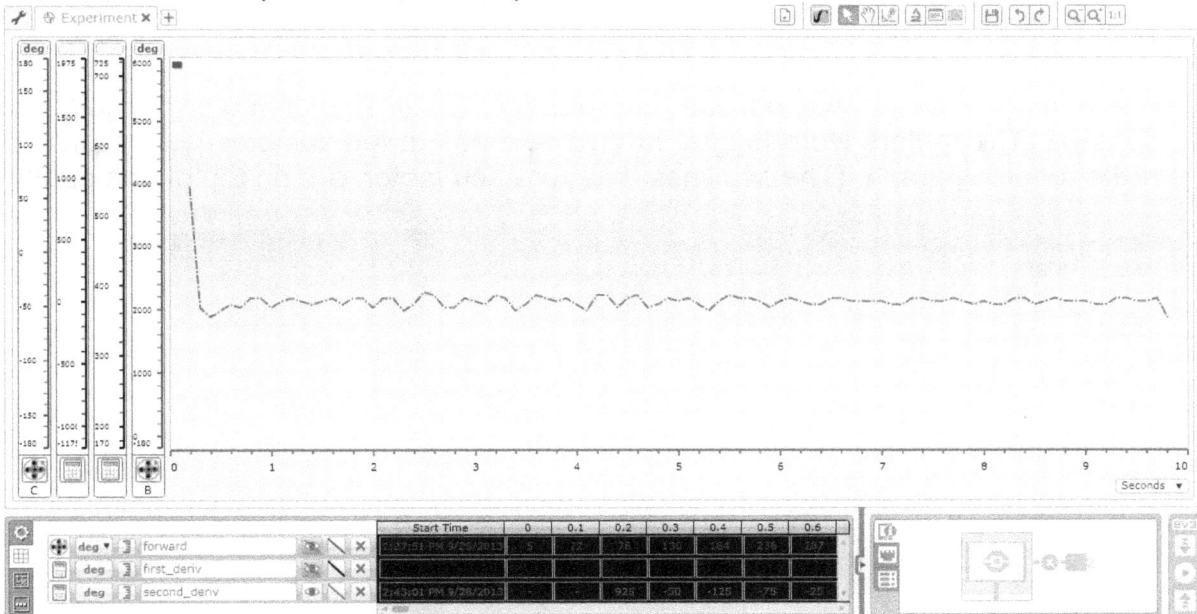

1. The velocity increased for the first fifth of a second, and then was constant. The line for the first half-second slopes upward (increasing speed), while after that the line is horizontal (steady speed).
2. The average velocity is around 510 degrees per second, or about 1.4 rotations/second.
3. The acceleration was positive, though decreasing, for the first fifth of a second and then zero after that. The car sped up during the first half-second (non-zero values) and then traveled at a fairly constant speed (the line bobbled around an acceleration of zero).
4. The overall acceleration is around zero.

Zigzag and Diamond

A two-motor EV3 car was run with the rotation sensor of each motor being logged in degrees. The motors were logged for five seconds at ten samples per second. Here is the resulting graph. (The two lines, one for each motor, are on top of one another.)

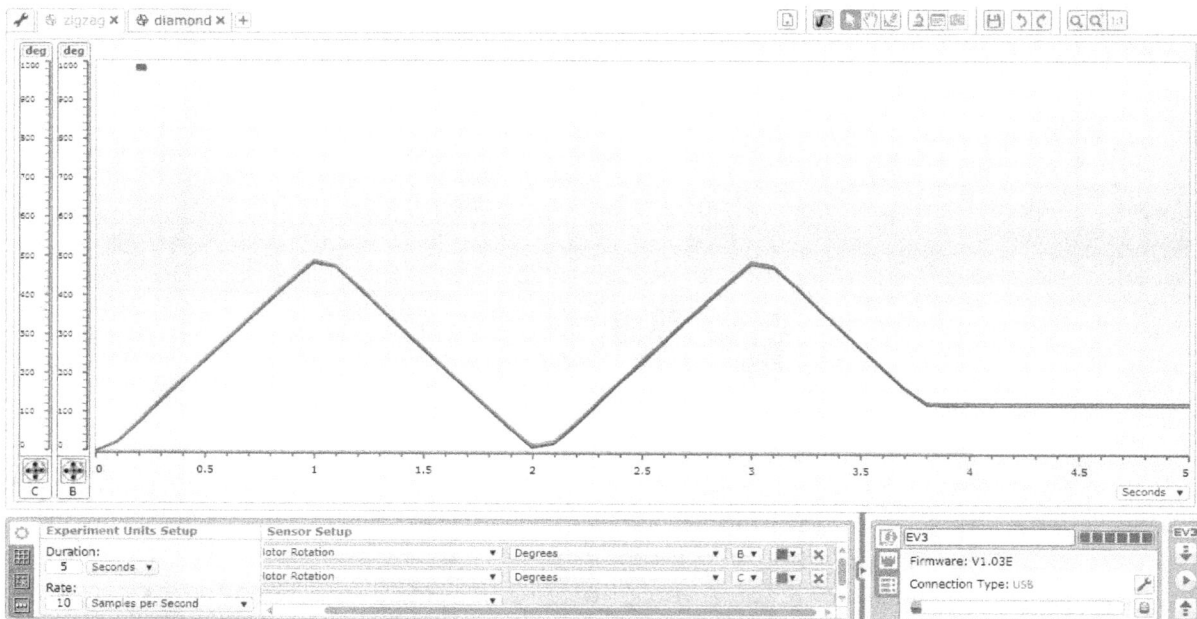

1. How do you think the car was moving during each section of the graph?
    a. First section, uphill line:

    b. Second section, downhill line:

    c. Third section, uphill line:

    d. Fourth section, downhill line:

2. Now try programming your car to mimic this graph. The picture above will help you set the correct parameters for logging (though you will need to figure out the Graph Program on your own).
    Notice that the graph axes have been set with a minimum of zero and a maximum of 1000. To change an axis value, click on it and type in a new number.
    How close did your graph come to matching the picture?

3. Modify your program until the graphs match. Describe your final program.

The car was run again, producing the following graph:

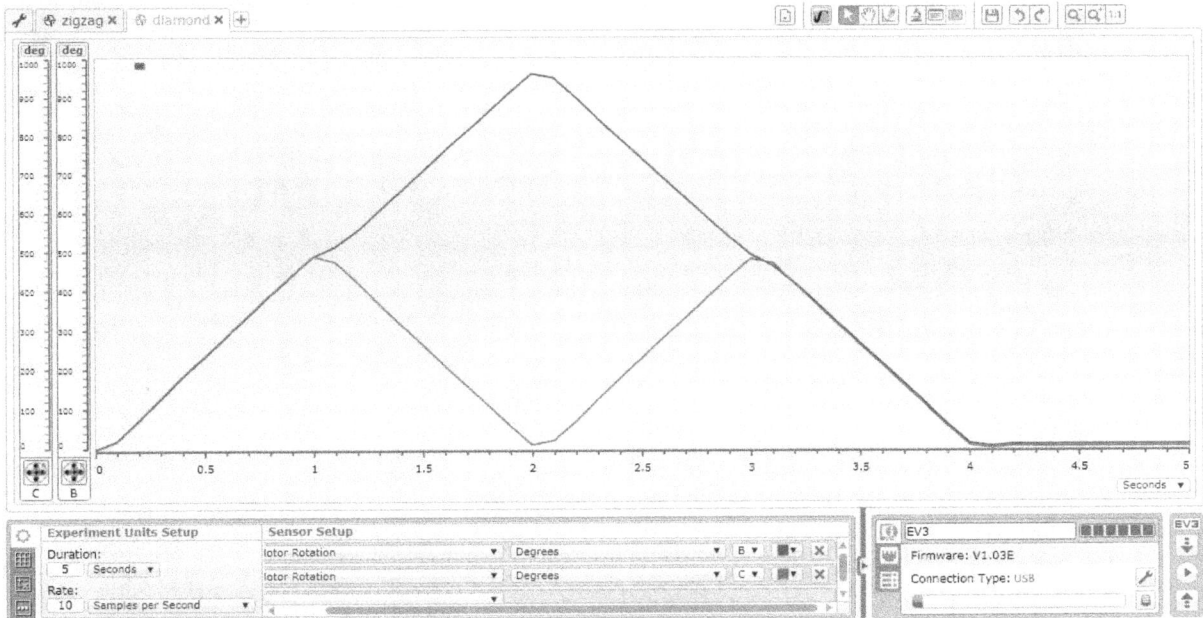

4. How do you think the car was moving during each section of this graph?
    a. First section, uphill line:

    b. Second section, first half of diamond:

    c. Third section, second half of diamond:

    d. Fourth section, downhill line:

5. Write a program to mimic this graph. Run your program and modify it if necessary to match the graph. Describe your final program.

Teacher Information

Zigzag and Diamond

In this activity, the students analyze rotation-sensor graphs, working backward to generate the motion of the car from the graph.

Objectives
1. To use a rotation sensor and data logging to analyze motion.
2. To interpret and reproduce a graph.

Materials
EV3 car, computer

Time: Approximately 40 minutes

Notes
1. Depending upon how the motors are oriented, the increasing slope may correspond to the actual car moving forward or moving backward. On the answer sheet, an uphill slope corresponds to forward motion.
2. As an extension, you can have the students write their own programs to generate mystery graphs, and then have them try to solve their classmates' graphs.

Sample Program for Zigzag and Diamond

Data Logging:

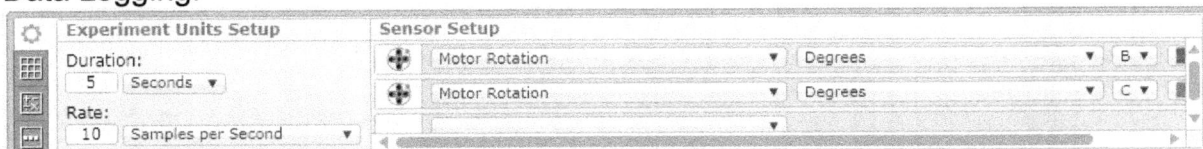

Zigzag Graph Program:

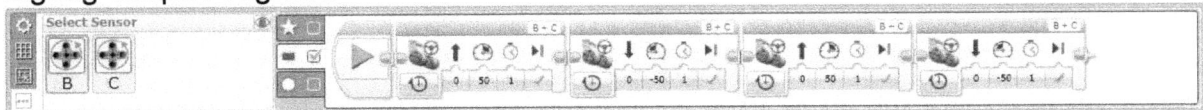

Diamond Graph Program:

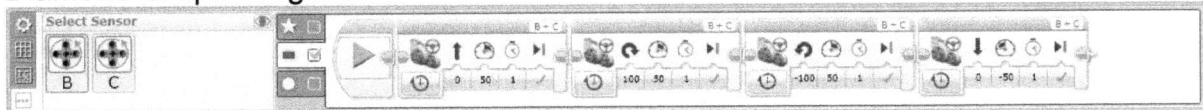

Answers to Zigzag and Diamond

1. The zigzag pattern, section by section. (Note: depending upon the orientation of the car's motors, the directions may be reversed; in other words, the car would move forward first, then backward.)
   a. First section, downhill line: Car drives backward for two seconds.
   b. Second section, uphill line: Car drives forward for two seconds.
   c. Third section, downhill line: Car drives backward for two seconds.
   d. Fourth section, uphill line: Car drives forward for two seconds.

2-3. Initial answers may vary. The final program should be similar to the one below.

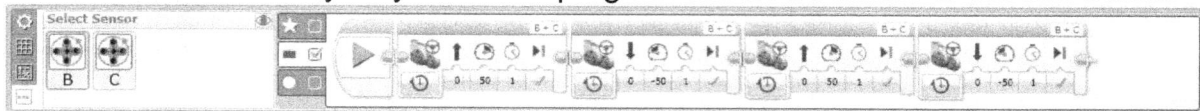

4. The diamond pattern, section by section. (Note: depending upon the orientation of the car's motors, the directions may be reversed; in other words, the car would move forward first, then backward.)
   a. First section, downhill line: Car drives backward for two seconds.
   b. Second section, first half of diamond: Car turns counterclockwise for two seconds.
   c. Third section, second half of diamond: Car turns clockwise for two seconds.
   d. Fourth section, uphill line: Car drives forward for two seconds.

5-6. Initial answers may vary. The final program should be similar to the one below.

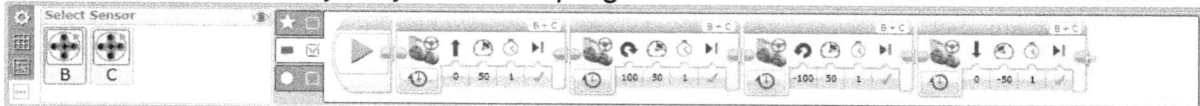

Puppy Data Logging

Add data-logging blocks to the beginning and end of your proportional puppy program to collect ultrasonic data as the puppy moves. Set the loop to run for ten seconds.

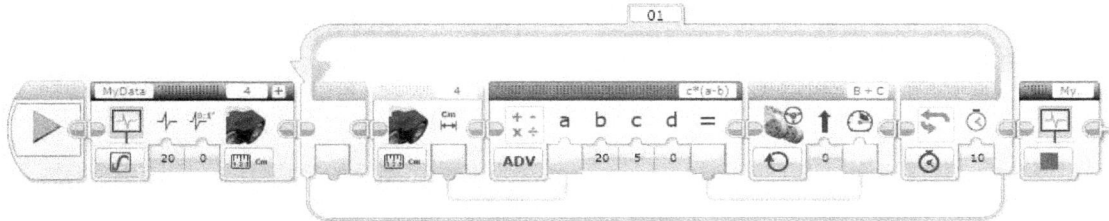

Run the puppy towards a wall, testing the different values of c, the gain, shown in the chart below. Make sure that you start the puppy the same distance from the wall each time, so that you will be able to compare the data easily.

To upload your data, click on the plus sign on the program bar at the top of the page and choose New Experiment. Once you upload your data, you may want to change the y-axis of the graph to make the details easier to see.

Value of c	Description of behavior	Sketch of graph
1		
5		
20		
50		

Teacher Information   Puppy Data Logging

The students investigate a feedback control system using data logging.

Objectives
1. To use data logging to investigate a feedback control system.
2. To see how changes in the feedback gain affect the outcome.

Materials
EV3, motors, ultrasonic sensor, LEGO® pieces including wheels, computer

Time: Approximately 30 minutes

Notes
1. This activity is designed as an extension of Proportional Puppy in the STEM Activities section, which should be completed first.
2. In order to see the oscillations clearly, you may need to have the students narrow the range of the y-axes of their graphs. To do so, click on the top and bottom numbers and type in the new numbers.
3. The answers will vary, depending upon the design of the car. Generally, the results should be similar to the answers outlined below.
4. If their puppy goes backward, the students will need to adjust their program to multiply the results of their Math block by negative one. An easy way to accomplish this is to reverse the subtraction, using b − a instead of a − b.

Sample Program for Puppy Data Logging

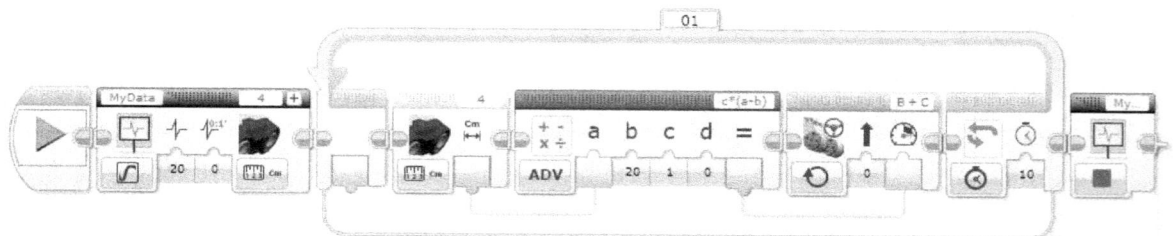

Answers to Puppy Data Logging

Value of c	Description of behavior	Sketch of graph
1	The car slows and stops before reaching the correct point.	
5	The car slows as it nears the correct point, stopping at or near the point with little or no oscillating.	
20	The car oscillates briefly, converging to the correct stopping point.	
50	The car goes unstable, oscillating around the correct stopping point.	

Which Room?

Part I: Conditional Probability

Imagine that you are a robot. You are in one of two possible rooms—one with all black walls and the other with two black walls and two white walls. You have an equal chance of being in each room; that is, the probability that you are in room 1 = ½.

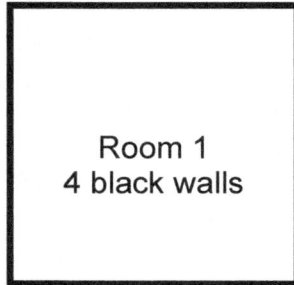

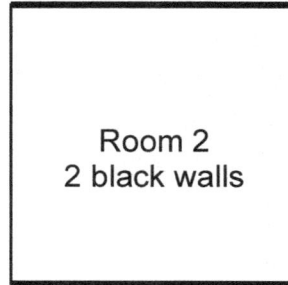

Now, with your color sensor, you detect that you are facing a black wall. With this new piece of information, what are the chances that you are in Room 1, the all-black room? Clearly, the probability is now greater than ½. But, how much greater?

To find out, we will use a type of mathematics called conditional probability. It turns out that the probability that you are in Room 2, given that you know one of the walls is black is

$$p(A|B) = \frac{p(B|A) \times p(A)}{p(B)}.$$

where A = you are in Room 1
    B = you see a black wall
    p() = the probability that the event in parentheses occurs.

So, let's put the equation above into words for this example:

$p(A|B)$ = the probability that you are in Room 1 given that you see a black wall—this is what we are trying to find out.

$p(B|A)$ = the probability that you see a black wall, given that you are in room 1. This is easy: if you are in Room 1, the probability that you see a black wall = 1. It is a sure thing!

$p(A)$ = the probability that you are in Room 1 without knowing anything else. This probability = ½, since there are two rooms.

p(B) = the probability that you will see a black wall. Since there are a total of eight walls and six of them are black, p(B) =6 out of 8 or ¾.

So, p(A|B) = $\dfrac{p(B|A) \times p(A)}{p(B)}$ = $\dfrac{1 \times ½}{¾}$ = $\dfrac{4}{6}$ or $\dfrac{2}{3}$.

1. How many black walls would you need to see to be absolutely sure (p = 1) that you are in Room 1 instead of Room 2?

Suppose we add a third room. Room 3 has three black walls and one white wall. Again, the robot detects a black wall.

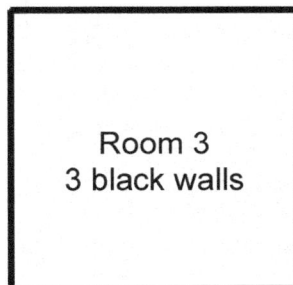

Room 3
3 black walls

2. Now what are the chances that the robot is in room 1? Are your chances better, worse, or the same as in the two-room situation? Show your work.

Part II: The Mystery Room

Now, you are ready to try your robot. Your robot will be placed in one of the four rooms shown below. You will be allowed to take five seconds of color-sensor readings while standing still.

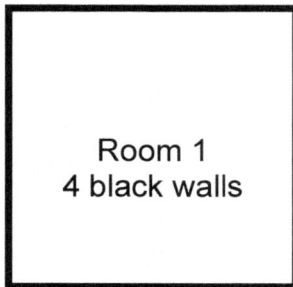

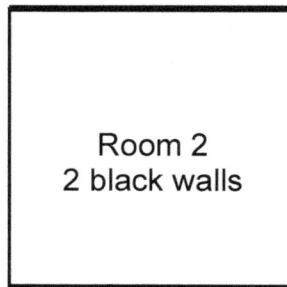

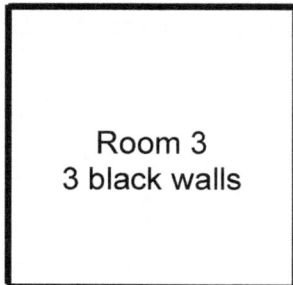

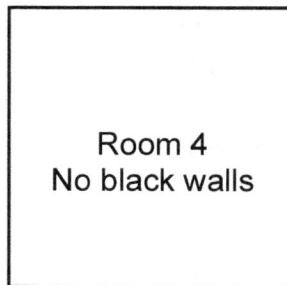

Room 1
4 black walls

Room 2
2 black walls

Room 3
3 black walls

Room 4
No black walls

Write a program to collect the data. Since you do not know how close to the wall your sensor will be, it is more useful to collect reflected light intensity data rather than color data.

Before trying your program in the mystery box, use the test box to calibrate your sensor so that you know what sensor readings correspond to black walls and white walls.

    3. Black wall reading in test box _____

    4. White wall reading in test box _____

Give your EV3 to the teacher, who will run your program in the mystery box and return the car to you.

    5. Color-sensor reading in mystery box _____

    6. Most likely color of wall _____

How many black walls is the mystery room most likely to have? To find out, calculate the conditional probability for Rooms 1-4 based upon your color-sensor reading. Show your work below.

7. Conditional probability that Room 1 is the mystery room:

8. Conditional probability that Room 2 is the mystery room:

9. Conditional probability that Room 3 is the mystery room:

10. Conditional probability that Room 4 is the mystery room:

Of course, usually a robot can take more than one reading in order to determine which room it is in. Suppose that you have fifteen seconds to take readings in the mystery room. Think about a strategy you might use.

Write a program to collect color-sensor data in the mystery box. You may move around inside the box and take as many readings as you like, but your program may not run for more than fifteen seconds.

11. Write a brief description of your program:

Give your EV3 to the teacher, who will run your program in the mystery box and return the car to you.

12. Describe the data you collected:

13. What do you think the inside of the mystery box looked like?

14. How confident are you of your results? Explain.

Teacher Information   Which Room?

The students investigate conditional probability using Bayes' Rule.

Objectives
1. To apply Bayes' rule to solve a few simple discrete-probability problems.
2. To program a robot to collect color-sensor data.
3. To design a plan for determining which of four rooms the robot is in.

Materials
EV3, color sensor, LEGO® pieces, cardboard "rooms" with black and white walls, computer

Time: Approximately 60 minutes

Notes
1. The students will need to know some very basic probability before beginning this lab. However, they do not need to know any conditional probability beforehand.
2. The mystery rooms can be constructed out of sheets of stiff black and white cardboard taped together at the corners. The advantage of using separate walls is that the number of black and white walls in each room can be easily changed.

Sample Programs for Which Room?

Program to take five seconds of color-sensor data in the mystery room:

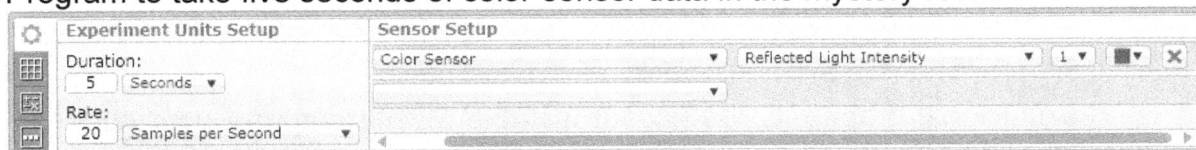

Program to take data while moving:

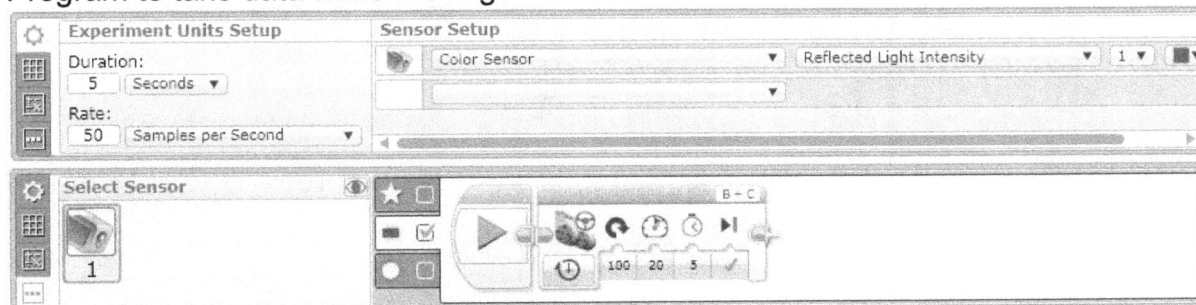

This program takes data while doing one slow 360-degree rotation.

Answers to Which Room?

Part I: Conditional Probability

1. You would need to see three black walls. As soon as you see more that two walls, you can be certain that you are not in Room 2, since it only has two black walls.

2. $p(A|B) = \dfrac{p(B|A) \times p(A)}{p(B)} = \dfrac{1 \times 1/3}{9/12} = \dfrac{4}{9.}$

    Your chances of being in Room 1 are lower.

Part II: The Mystery Room

3-6. Answers will vary.

7. Conditional probability that Room 1 is the mystery room if your color-sensor reading indicated a black wall:

    $p(A|B) = \dfrac{p(B|A) \times p(A)}{p(B)} = \dfrac{1 \times 1/4}{9/16} = \dfrac{4}{9.}$

    If your color-sensor reading indicated a white wall:

    $p(A|B) = \dfrac{p(B|A) \times p(A)}{p(B)} = \dfrac{0 \times 1/4}{7/16} = 0.$

8. Conditional probability that Room 2 is the mystery room if your color-sensor reading indicated a black wall:

    $p(A|B) = \dfrac{p(B|A) \times p(A)}{p(B)} = \dfrac{1/2 \times 1/4}{9/16} = \dfrac{2}{9.}$

    If your color-sensor reading indicated a white wall:

    $p(A|B) = \dfrac{p(B|A) \times p(A)}{p(B)} = \dfrac{1/2 \times 1/4}{7/16} = \dfrac{2}{7.}$

9. Conditional probability that Room 3 is the mystery room if your color-sensor reading indicated a black wall:

    $p(A|B) = \dfrac{p(B|A) \times p(A)}{p(B)} = \dfrac{3/4 \times 1/4}{9/16} = \dfrac{1}{3.}$

    If your color-sensor reading indicated a white wall:

    $p(A|B) = \dfrac{p(B|A) \times p(A)}{p(B)} = \dfrac{1/4 \times 1/4}{7/16} = \dfrac{1}{7.}$

10. Conditional probability that Room 4 is the mystery room if your color-sensor reading indicated a black wall:

    $p(A|B) = \dfrac{p(B|A) \times p(A)}{p(B)} = \dfrac{0 \times 1/4}{9/16} = 0.$

    If your color-sensor reading indicated a white wall:

    $p(A|B) = \dfrac{p(B|A) \times p(A)}{p(B)} = \dfrac{1 \times 1/4}{7/16} = \dfrac{4}{7.}$

11. Answers will vary. The students may collect as many readings as possible while rotating the robot to see multiple walls. Alternatively, they may program the robot to turn exactly 360 degrees while collecting readings.
12. Answers will vary.
13. Answers will vary depending upon the configuration of the mystery box and the quality of the students' data.
14. The students should not be expected to be able to generate the conditional probability of multiple events. However, they should have a general sense of how compelling (or not) their data are.

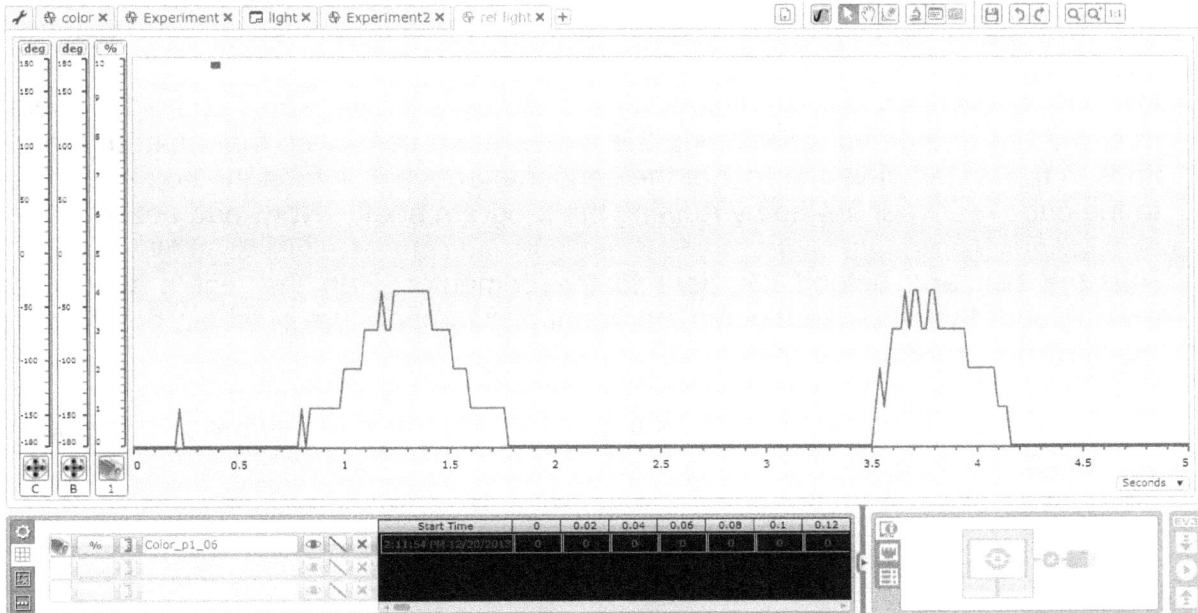

Sample data in a room with alternating black and white walls. The EV3 is timed to make one complete revolution, taking color-sensor data as it spins.

Stir It Up

People stir hot drinks to cool them down. Does stirring really help? Find out, using an EV3, a motorized stirrer, and a temperature sensor.

Build a LEGO stirrer. To construct it, attach a long axle to a motor. Next, design a part that can be attached to the axle to stir the water when the motor is turned on.

For the stirring program, log the temperature sensor in degrees Celsius once every five seconds for 500 seconds. Run the stirrer motor at full power.

Use LEGO pieces to build a holder for the temperature sensor. The holder should fit over the rim of the cup, positioning the temperature sensor so that most of the metal tip is in the water while the rest of the sensor is above it. Attach the holder and stirrer to the cup. Test your set-up by running the program briefly. Now, add hot liquid.

Run the test and upload the data to the computer. Run the test a second time, making sure that you use the same amount and temperature of water, but do not turn the stirrer on. Upload the data.

1. Why is it important to run the test twice, once with the stirrer on and once with it off?

2. Based upon your results, does stirring cool the liquid down faster?

Teacher Information   Stir It Up

In this activity, the students build motorized stirrers and use them to test whether stirring a hot drink speeds cooling.

Objectives
1. To use a computer to collect, graph, and analyze data.
2. To understand the importance of having a control when running an experiment.
3. To understand how heat energy can be dissipated.

Materials
EV3, temperature sensor, motor, LEGO® parts, beaker or mug, hot water, computer

Time: Approximately 40 minutes

Notes
1. This lab makes use of the NXT temperature sensor, which can be purchased separately.
2. This activity and the next two are related. The students use what they learn in the first two activities to design and build a cooling device in the third.
3. This activity is similar in many ways to the next one, It's a Breeze. The students can do both to gain practice in logging and analyzing data. Alternatively, half the class can do Stir It Up while the other half does It's a Breeze. The groups can then share their results and all of the students can apply what they've learned in Cool It Fast.
4. Most students will be surprised to discover that stirring does not help to cool the hot liquid. This activity contrasts nicely with It's a Breeze, where students discover that blowing on hot liquid does help to cool it.
5. Make sure that the students understand the importance of a control in their experiment. All of the conditions for the stirrer and control containers should be the same except for the stirrer—same amount of water, same initial temperature, etc. If two temperature sensors per group are available, the experimental and control set-ups can be run simultaneously. Otherwise, they can be run serially or a single control can be run for the entire class..
6. This activity can be used to discuss heat transfer. Heat is transferred to the cup and to the boundary layer of air adjacent to the liquid by conduction. The warmer air then rises, carrying heat away with it (convection). Does stirring help speed this process? Not in this case. Stirring increases the movement of liquid within the cup, but the temperature within a small cup is already relatively uniform. Thus, bringing liquid from the center of the cup to the surface does not enhance the cooling. (If you were doing this experiment using a large vat of hot liquid, then stirring would help, because the temperature of the liquid would not be uniform throughout the vat and stirring would help to mix it.)
7. Be sure to keep the liquid away from the motor and the EV3.

8. A stable, heat-resistant container, such as a 250-ml glass beaker or coffee mug, works best for this experiment.

Sample Program for Stir It Up

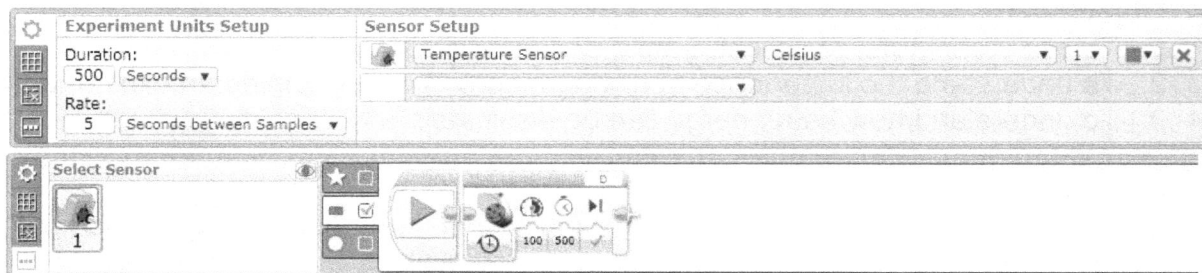

Answers to Stir It Up

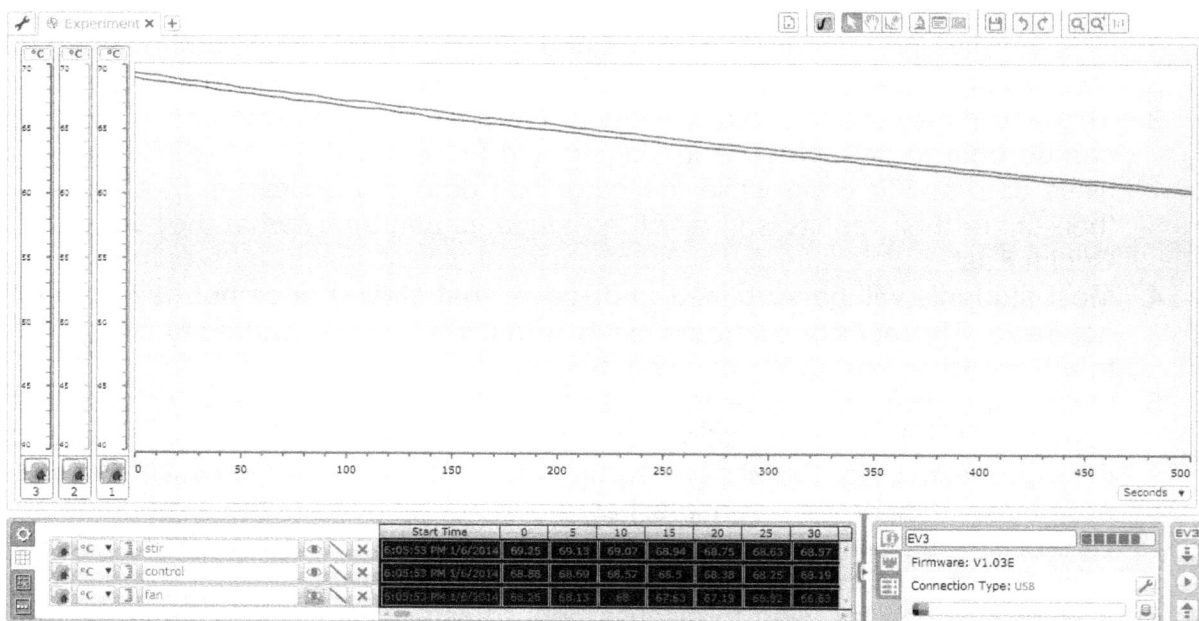

Sample data using 200 ml of water: Stirring makes no discernible difference in the cooling rate. The two curves have the same slope.

1. It is important to run the test twice, once with the stirrer on and once with it off, because the cup without the stirrer serves as a control. It gives a basis for comparison, so that one can tell if stirring made the temperature drop faster than it would have in an undisturbed cup.
2. No, stirring appears to make little or no difference in cooling the liquid. (See the Teacher Information page for an explanation.)

It's a Breeze

People blow on hot drinks to cool them down. Does blowing really help? Find out, using an EV3, a temperature sensor, and a fan.

For the program, log the temperature sensor in degrees Celsius once every five seconds for 500 seconds.

Use a portable fan to create the breeze for this activity. Position the fan and the cup so that the breeze blows across the surface of the hot liquid.

Use LEGO pieces to build a holder for the temperature sensor. The holder should fit over the rim of the cup, positioning the temperature sensor so that most of the metal tip is in the water while the rest of the sensor is above it. Attach the holder to the cup and add hot liquid to the cup.

Run the test and upload the data to the computer. Run the test a second time, making sure that you use the same amount and temperature of water, but do not turn the fan on. Upload the data.

1. Why is it important to run the test twice, once with the fan on and once with it off?

2. Based upon your results, does blowing cool the liquid down faster?

Teacher Information   It's a Breeze

In this activity, the students use a portable fan to test whether or not blowing on a hot drink speeds cooling.

Objectives
1. To use a computer to collect, graph, and analyze data.
2. To understand the importance of having a control when running an experiment.
3. To understand how heat energy can be dissipated.

Materials
EV3, temperature sensor, portable fan, beaker or mug, hot water, computer

Time: Approximately 30 minutes

Notes
1. This lab makes use of the NXT temperature sensor, which can be purchased separately.
2. This activity is the second of three related experiments. The students use what they learn in the first two to design and build a cooling device in the third.
3. This activity is similar in many ways to the previous one, Stir It up. The students can do both to gain practice in logging and analyzing data. Alternatively, half the class can do Stir It Up while the other half does It's a Breeze. The groups can then share their results and all of the students can apply what they've learned in Cool It Fast.
4. The students will discover that blowing does help to cool the hot liquid. This activity contrasts nicely with Stir It Up, where the students discover that stirring a hot liquid does not help to cool it.
5. Make sure that the students understand the importance of a control in their experiment. All of the conditions for the experimental and control containers should be the same except for the fan—same amount of water, same initial temperature, etc. If two temperature sensors per group are available, the actual and control experiments can be run simultaneously. Otherwise, they can be run serially or a single control can be run for the entire class.
6. This activity can be used to discuss heat transfer by convection. In convection, heat is transferred by the movement of a heated substance itself, such as the blowing of heated air. When a cup of hot liquid cools, convection currents carry the heated air upward, since the heated air is less dense. The air currents produced by the fan accelerate this process, carrying the heated air away at a much faster pace, allowing cooler air to take its place and be heated by the liquid.
7. Be sure to keep the liquid away from the motor and the EV3.
8. This activity works well with a 250-ml beaker or standard-sized coffee cup filled almost to the brim.

Sample Program for It's a Breeze

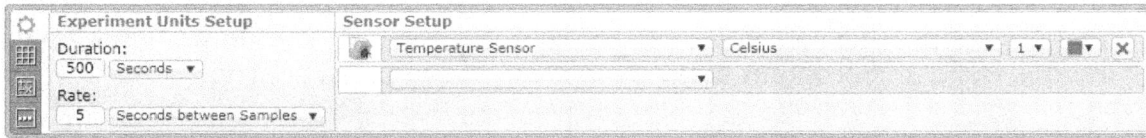

Answers to It's a Breeze

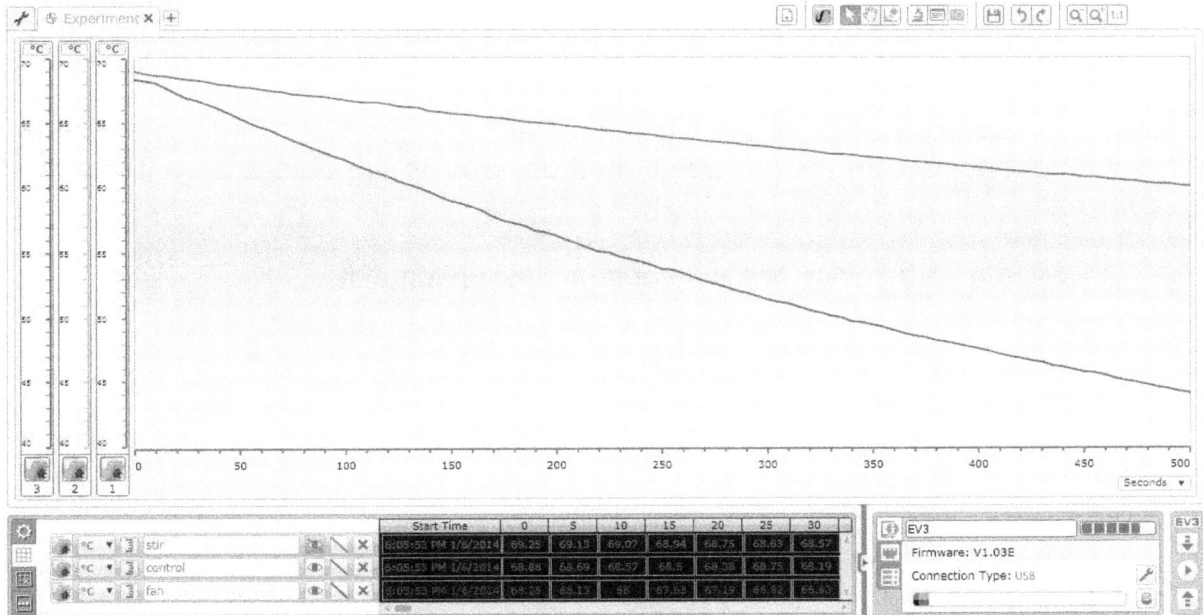

Sample data: The difference between the top curve, the control, and the bottom curve, cooled with a fan, shows that the fan increases the rate of cooling.

1. It is important to run the test twice, once with the fan on and once with it off, because the test without the fan serves as a control. It gives a basis for comparison, so that one can tell if the moving air caused the temperature to drop faster than it would have in an undisturbed cup.
2. Yes, the fan increases the cooling rate significantly. (See the Teacher Information page for an explanation.)

Cool It Fast

You are impatient to drink your cocoa, but it is too hot. You do not want to wait for it to cool. Build and program a device that will speed up its cooling.

Build a stirrer, a fan, or your own idea for a cooling device.

Write a program to run your device while collecting temperature data every five seconds for 500 seconds.

Set up your device to cool a cup of hot liquid. Remember to position the temperature sensor inside the cup so that the metal tip will be submerged, but the rest of the sensor will not.

Run your program and upload the data. Refill the cup with hot liquid and run the experiment again with your device turned off. Upload the data and compare the two data sets.

1. Sketch your device and write a brief description of it.

2. Did your device cool the liquid more quickly than in the control cup?

3. If you had more time, how would you improve your device?

Teacher Information   Cool It Fast

In this activity, the students build and test their own devices to speed the cooling of a hot drink.

Objectives
1. To use a computer to collect, graph, and analyze data.
2. To understand the importance of having a control, or other meaningful basis for comparison, when running an experiment.
3. To understand how heat energy can be dissipated.
4. To make use of what has been learned to design a cooling device.

Materials
EV3, temperature sensor, motor, LEGO® parts, beaker or mug, hot water, computer

Time: Approximately 60 minutes

Notes
1. This lab makes use of the NXT temperature sensor, which may be purchased separately.
2. This activity is the third of three related experiments. The students use what they learned in the first two to design, build, and test a cooling device in the third.
3. This activity can be used to assess the students' understanding of heat transfer, because they are using what they have learned to design an effective cooling device.
4. Based upon the previous labs, most students will probably choose to build fans. Many students will gear up their fans in order to increase the rotation speed of the blades.
5. Make sure that the students understand the importance of a control in the design of their experiment. All of the conditions for the experimental and control set-ups should be the same except for the cooling device—same amount of water, same initial temperature, etc. If you wish, you can run one control for the entire class, as long as everyone agrees upon the initial conditions.
6. If the students wish to compare their devices with each other, make sure that they understand the importance of keeping all of the other factors the same. If the initial conditions differ, then the comparison is meaningless.
7. Be sure to keep the liquid away from the motors and the EV3.

Engineering Projects

These engineering projects give the students a chance use the engineering design process, as well as their creativity, to find solutions to challenges. In doing the projects, they practice the building and programming skills they have learned.

Different activities emphasize different aspects of engineering design: brainstorming, prototypes, design criteria, and so on. Notes about the focus of each project can be found on the Teacher's Information page for that activity.

These projects can be fairly simple or impressively elaborate, depending upon the time available. Many of the students enjoy the creative license that projects like these afford them. If possible, arrange a way for the students to share their creations with an audience, perhaps a class of younger students.

The activities in this section are:
1. Meet and Greet
2. Fairytale Fix: Goldilocks
3. Fairytale Fix: Rapunzel
4. Fairytale Fix: Cinderella
5. Dancing Bot
6. Household Helper
7. Music Box
8. Mini Golf
9. Robotic Zoo
10. Chain Reaction Machine
11. EGGcellent Contraption
12. Wacky Gumball Machine
13. Robo Artist

Meet and Greet

Build a robot that is the life of the party. When someone greets it—by shaking its hand or waving to it (your choice), the robot responds with enthusiasm. It can move, make sounds, display messages, whatever you wish. The more effusive, the better.

1. Brainstorming: What sensors could you use for this activity? How would you use them? List three possibilities and explain how you might use each one.

   a.

   b.

   c.

2. What does your robot look like? Draw a sketch and write a brief description. Your sketch should be detailed enough that a visitor could use it to pick out your robot from all of the others.

3. What does the robot do when you greet it?

Teacher Information   Meet and Greet

The students build robotic pals who greet newcomers with enthusiasm.

Objectives
1. To use the engineering design process to craft a creative solution to a challenge.
2. To gain additional practice using sensors.

Materials
EV3, motors, sensors, LEGO® pieces, computer, craft materials (optional)

Time: Approximately 60 minutes

Notes
1. This activity is both straightforward and open-ended, with a large range of possible solutions. It is a good design project for beginners. It can be done by students after they have completed the first programming sequence, Seeing Red.
2. You may want to discuss the engineering design process at the beginning of this activity.
3. This activity introduces the concept of brainstorming.
4. You may want to talk about effective sketching techniques—views from different angles, close-up views of some features, labeled parts—before the students sketch their robots.
5. The features of the pal can be done with LEGO pieces, craft materials, or both.

Answers to Meet and Greet

1. A variety of sensors are possible. If the students choose to have the pal activated by a handshake, they can use a gyro sensor or touch sensor. For a wave-activated pal, a light sensor or ultrasonic sensor can be used instead.
2-3. Answers will vary.

Fairytale Fix: Goldilocks

Goldilocks sat on Baby Bear's chair, breaking it and making Baby Bear very sad. To give this fairytale a happier ending, build a better chair for Baby Bear. Make it sturdy enough that Goldilocks will not break it. Better yet, design it so that Goldilocks is discouraged from sitting in it in the first place. An alarm that goes off if she tries to sit down, perhaps?

Build a Solution
Build a chair for the bear using LEGO pieces. You can include the EV3 as part of the chair or not—your choice.

Program a Solution
Attach an alarm to the chair to keep intruders away, one that Baby Bear can set before he leaves the house. The type of sensor is up to you—touch, color, ultrasonic—whatever you like.

Some additional features to consider adding:
- An alarm that includes a warning message as well as sound.
- An alarm that repeats the warning for a set amount of time.
- An alarm that resets itself so that it can be used again.
- A chair that ejects the intruder.
- A special feature of your own design.

Thanks to you, Baby Bear can enjoy his walk in the woods knowing that his sturdy chair will be waiting for him, safe and sound, when he returns.

Fairytale Fix: Rapunzel

Rapunzel was locked in a tall tower by a witch, with no way for anyone to enter except by climbing Rapunzel's long hair up the outside of the tower to the window. The prince snuck up to see her by calling "Rapunzel, Rapunzel, let down your hair, that I may climb the golden stair. This sounds painful for Rapunzel!

Build a Solution
Build a device that Rapunzel can lower out the tower window and then pull up with a person on board.

To help you, here are some mechanisms for raising and lowering objects:

Winch: A winch is a device for winding and unwinding string. A simple winch can be just a wheel hub with a crank handle.

Rachet and pawl: A ratchet and pawl lets a toothed wheel (the ratchet) move in one direction only. As it turns, the pawl meshes with its teeth, preventing the wheel from turning backwards. Hint: If you were winding up a string on a spool, including a ratchet and pawl would keep the spool from reversing direction and letting the string unwind.

Here are some additional building features you might include:
- A platform for the rider to stand on or a chair to sit in.
- A safety mechanism to prevent the rope from suddenly unwinding while it is lifting someone.
- A device that can be folded up or disguised as something else, so that Rapunzel can keep it a secret from the witch.
- Your own special feature.

Program a Solution
Lifting a witch—or a prince—is hard work. Motorize Rapunzel's contraption so that she doesn't need to do the heavy lifting by hand.

Design a control for the device. Here are some possibilities:
- A lift that goes up when you say "go" and halts when you say "stop."
- A device that rises while a button is being pushed and stops while the button is released.
- A control that can use the motor to raise *or* lower the device.

Rapunzel thanks you for your help in sparing her hair.

Fairytale Fix: Cinderella

Poor Cinderella must leave the ball by midnight, when her dress will become rags and her coach will become a pumpkin. Unless you can perform magic, you can't modify the spell. However, you can help her out by upgrading the coach. Build and program a motorized "smart coach" to help her escape before midnight.

Build a Solution

Build a motorized EV3 coach for Cinderella, so that she does not need to rely on mice that have been turned into horses. The coach should contain a seat that holds the passenger securely while the coach is moving.

Program a Solution

Make a smart coach.

Some possible "smart" features to consider adding:

- The coach includes a warning signal that goes off a few minutes before midnight.
- The coach moves faster the later it gets.
- The coach moves faster the darker it gets.
- The coach includes start and stop buttons or a steering mechanism.
- The coach discourages or disables pursuers in some way.

Thanks to you, Cinderella can make a clean getaway from the ball.

Teacher Information   Fairytale Fixes

The students engineer solutions to fairytale dilemmas.

Objectives
1. To explore the idea of engineers as creative problem solvers.
2. To plan and carry out a design project.

Materials
EV3, motors, sensors, LEGO® pieces, minifigures, computer

Time: Approximately 60 minutes each

Notes
1. This set of activities provides more scaffolding than some of the following more open-ended projects. For this reason, these fairytale fixes make good introductions to engineering design.
2. You may want to discuss the engineering design process at the beginning of this activity.
3. The three fairytale fixes are listed in approximate order of complexity, though there can be considerable variation in the difficulty of each one, depending upon the fixes decided upon by the students.
4. You can assign one, two or all of the fairytale fixes to every student or present all three scenarios and let the students choose one.
5. As a follow-up activity, you can have the students choose their own fairytales, fables, or familiar stories to fix.

Dancing Bot

Teach your EV3 to dance.

Design criteria:
- Your dance program must be at least 30 seconds long.
- It must include at least three different moves (for example, move forward, spin to the left, move backward) and at least one repeated element.
- Your EV3 dancer must use at least two motors.
- Your dancer must have a face. It can be human, animal, robot, space creature, whatever you want—as long as it has a face.

1. On a separate sheet, make a labeled sketch of your dancer. Indicate the positions of the motors.

2. Describe your dance.

3. Describe a difficulty you encountered in building and programming your dancer. How did you address the problem?

4. If you had more time, what changes would you make or what features would you add?

Teacher Information   Dancing Bot

The students write their own programs to create dancing robots.

Objectives
1. To create a design that meets a set of criteria.
2. To gain additional programming practice.

Materials
EV3, motors, LEGO® pieces, computer, craft materials (optional)

Time: Approximately 50 minutes

Notes
1. This activity introduces the idea of design criteria—requirements that the finished robot must fulfill.
2. Any simple two-motor robot can be used for this activity. You may want to suggest EV3 configurations to the students, such as an upright or slanted EV3.
3. The students can use the medium motor to create additional movement, such as waving arms.
4. Because at least thirty seconds of music are needed for this activity, it is better to use a computer or other outside source to play the music. (The EV3 Sound Editor will only let you save five seconds or fewer of recorded sound.) You can choose one song for the entire class to use or let each group choose its own song.

Household Helper

Every home could use a smart appliance, gadget, or piece of furniture. Think how handy it would be to have a bed that ejects you if you oversleep. Or a refrigerator that suggests you eat an apple rather than a slice of cake when you open the door to get a snack.

Brainstorm ideas for smart furniture/appliances/gadgets. List at least three.

1.

2.

3.

Choose one of the ideas above and build a prototype for it. Designers build prototypes of their ideas to test them out and see whether they are feasible. Often, the prototype is smaller than the finished product will be.

4. Describe one aspect of your prototype that works well.

5. Describe one thing that you would change if you actually built your household helper.

6. Make a labeled sketch of your prototype on a separate sheet.

Teacher Information   Household Helper

In this activity, the students construct prototypes of smart appliances, household gadgets, or pieces of furniture.

Objectives
1. To brainstorm ideas based on a given theme.
2. To construct and evaluate a prototype.
3. To use the engineering design process to solve a problem.

Materials
EV3, motors, sensors, LEGO® pieces, computer, craft materials (optional)

Time: Several class periods

Notes
1. You may want to review the engineering design process as you introduce this activity.
2. This activity introduces the idea of prototyping. You may want to show the students examples of prototypes before starting this activity.
3. As an extension of this activity, you can have the students produce instruction manuals to explain how to use their devices or commercials to advertise them.

Music Box

Build a music box in which figures move while music plays. Your box must have a theme with appropriate music, figures, and decorations.

Your music box must include two or more figures that move in different ways. At least one must move back and forth or up and down.

1. Title of your music box:

2. Theme of your music box:

3. Music played:

4. Describe your figures/action/storyline:

5. Sketch and describe the mechanism you used to create lateral movement:

Creating Lateral Motion with the EV3 Motor:
Cams and Offset Gears

Using a cam to create up-and down motion.

Using an offset gear to create up-and down motion.

Creating Lateral Motion with the EV3 Motor:
Gear Racks and Four-bar Linkages

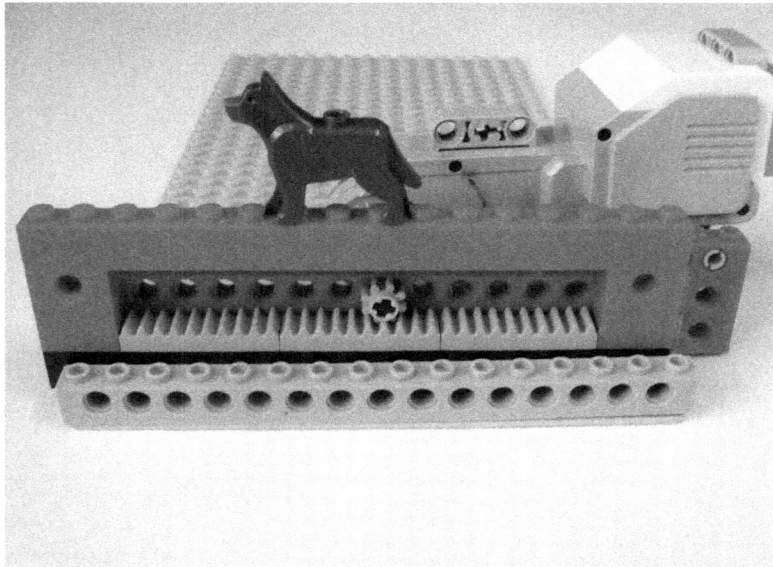

Using a gear rack to create back-and-forth motion.

Using a four-bar linkage to create back-and-forth motion.

Teacher Information   Music Box

The students build music boxes with themes of their choice.

Objectives
1. To use technical knowledge to solve a problem.
2. To create figures that move in different ways, including back-and-forth or up-and-down movement.
3. To write a program to control different movements at the same time.

Materials
EV3, motor, LEGO® pieces including minifigures, computer

Time: Several class periods

Notes
1. In this activity the students make use of technical knowledge to design a project by using motors in a variety of ways to produce different types of movement.
2. The students should already be proficient in using motors to produce rotary motion. The accompanying Creating Lateral Motion pages show some possibilities for creating back-and-forth or up-and-down movement.
3. The students can include the song as part of the EV3 program or play one separately. To write a simple tune, string together Sound blocks, using the play note option to set the pitch and duration of each note. It is also possible to record sound using the Sound Editor under Tools, but only recordings of five seconds or fewer can be saved.

Mini Golf

As a class, we will build a miniature golf course. The "golf ball" for our course will be a marble. Each group will be responsible for designing and constructing one hole. You may use LEGO pieces and any other materials you wish.

Design criteria:
- Your hole must include a starting tee of some type and an ending spot. The ball, if hit correctly from the starting tee, should traverse the path through your hole and finish at the ending spot.
- Your hole must use an EV3 and at least one sensor. Some action (movement, sound, lights, etc.) must happen in response to the ball.
- Your design must have a theme, such as outer space or penguins. Once you have decided on a theme, register it with the teacher. Themes are first come, first served—if some other group has already registered the brilliant idea that you planned to use, you will need to think of another theme.

Engineering notebook:
- For the duration of this project, you will keep an engineering notebook.
- After each work session, you must make an entry explaining your design thinking and progress. Your entry should be detailed, complete, and thoughtful. Include labeled sketches of the mechanisms involved.
- The engineering notebook documents your journey—include ideas that don't pan out and changes you make to your design.
- If you are absent, you still need to fill out a sheet, noting that you were absent and describing what your group did that day.

Good luck!

Teacher Information   Mini Golf

The class designs a mini-golf course, with each group designing and constructing one hole.

Objectives
1. To plan and carry out a major design project.
2. To keep an engineering notebook.

Materials
EV3, motors, sensors, LEGO® pieces, marbles or metal ball bearings, computer, craft materials (optional)

Time: Several class periods

Notes
1. This project can be done from a systems perspective, by emphasizing that the holes must all be done to the same scale and working to coordinate the holes so that they form a coherent sequence of play.
2. Keeping an engineering notebook helps the students organize their time and document the design process.
3. The decorations can be done using LEGO pieces, craft materials, or both.
4. This project can also be done on a larger scale with standard golf balls. If golf balls are used, have the students design only an obstacle around the ending spot, rather than the entire hole from tee to end.

Robotic Zoo

Build an animal for the robotic zoo. It can be a land animal, a sea creature, an insect, a bird—it does not have to be an animal that is traditionally found in a zoo.

Design criteria:
- The animal must move in some way. The movement can be walking, or it can be opening and closing jaws, wagging a tail--whatever makes sense for your animal. If your animal does several types of movement, even better.
- The animal must use at least one sensor and react to a change in its environment.

Engineering notebook:
- For the duration of this project, you will keep an engineering notebook.
- After each work session, you must make an entry explaining your design thinking and progress. Your entry should be detailed, complete, and thoughtful. Include labeled sketches of the mechanisms involved.
- The engineering notebook documents your journey—include ideas that don't pan out and changes you make to your design.
- If you are absent, you still need to fill out a sheet, noting that you were absent and describing what your group did that day.

Good luck!

Teacher Information   Robotic Zoo

The students design and construct robotic animals.

Objectives
1. To plan and carry out a major design project.
2. To keep an engineering notebook.

Materials
EV3, motors, sensors, LEGO® pieces, computer, craft materials (optional)

Time: Several class periods

Notes
1. This project can be done from a systems perspective, by emphasizing that the animals must all be done to the same scale. You may want to have the students organize the zoo into sections—land animals, birds, etc. The students can also work together to produce the additional infrastructure needed for a zoo—front entrance, snack bar, veterinary office, etc. One possibility is to have each group produce one animal and one support structure.
2. Keeping an engineering notebook helps the students organize their time and document the design process.
3. The decorations can be done using LEGO pieces, craft materials, or both.

Chain Reaction Machine

We will build a class contraption that carries a marble from one end to the other. Each group will design and build one section.

The order of the sections will be assigned. You will need to talk with the groups before and after you to find a mutually agreeable height for the ball's exit and entrance.

Here are the criteria for your section:

- It can be as tall as you want, but its footprint can be no larger than 60 cm long or wide.
- The marble must have a height change of at least ten centimeters in the contraption. If the marble enters and leaves at approximately the same height, it will need to go up and back down—or down and back up.
- The marble must take at least 15 seconds to traverse your section.
- Your section must contain at least one sensor and at least one simple machine.

Contraption Criteria	
Length and width no more that 60 cm	
Starting height matches section before	
Ending height matches section after	
10 cm height change	
Simple machine used	
Sensor used	
Marble takes 15 seconds in machine	
Marble traverses machine unaided	

Teacher Information   Chain Reaction Machine

The class designs a chain reaction machine, with each group designing and constructing one section.

Objectives
1. To plan and carry out a major design project.
2. To coordinate with other groups in planning and executing a design.

Materials
EV3, motors, sensors, LEGO® pieces, craft materials, ruler, stopwatch, computer, sheets of cardboard (optional)

Time: Several class periods

Notes
1. The nature of this project ensures that it will involve extensive testing and redesign. Allow plenty of time for testing and modifying the contraptions.
2. If you do not wish to have the groups coordinate the exits and entrances of their sections of the machine, you may assign a set entrance and exit height: Your contraption must accept the marble at a height of 10 cm and deliver it to the next contraption at the same height. The marble must be moving at a reasonable pace when it leaves your contraption.
3. You may want to introduce this project by showing the students some Rube Goldberg cartoons or videos.
4. Large LEGO baseplates or pieces of cardboard make good foundations for the contraptions.

EGGcellent Contraption

Build an egg-delivery contraption, which will deliver an egg safely from a nest on the table to a collecting plate on the floor.

The nest must contain a sensor that detects when an egg is laid in the nest and starts the contraption. The contraption removes the egg from the nest and delivers it to a plate on the floor below, without damaging the egg in any way. The contraption then resets itself, ready for another egg. The contraption must contain two simple machines. When not in use, the contraption must fit in your storage container. Your contraption may be made of LEGO pieces and/or other materials.

You may test your contraption as often as you wish using plastic eggs weighted with clay. When your contraption can consistently collect plastic eggs, you may try real eggs. Good luck!

Contraption Criteria	
Egg detected by sensor	
Egg removed from nest	
Egg transferred from table to floor	
Contraption contains a simple machine	
Contains a second simple machine	
Contraption resets after egg delivery	
Delivers two eggs in a row without damage	
Nest looks like a nest	
Contraption fits in storage container	

Teacher Information   EGGcellent Contraption

The students design and construct contraptions to move eggs safely from the table to the floor.

Objectives
1. To plan and carry out a major design project.
2. To engage in extensive testing and redesigning.

Materials
EV3, motors, sensors, LEGO® pieces, eggs, plastic eggs, clay for adding weight to plastic eggs, plates, string, craft materials, cleaning supplies (in case of broken eggs), computer

Time: Several class periods

Notes
1. The nature of this project ensures that it will involve testing and redesign. Allow plenty of time for testing and modifying the contraptions.
2. This project can be done from a systems perspective by breaking the project into sections. One group can build a nest, another group can build a device to remove the egg from the nest, a third group can build a device to transport the egg to the floor, and a fourth group can program the sensor and motors to control the various pieces of the project.
3. You may want to use hard-boiled eggs to minimize the mess and the risk of salmonella. If you use raw eggs, make sure to follow proper cleaning and hand-washing procedures.
4. The contraptions can be built using LEGO pieces, craft materials, or both.

Wacky Gumball Machine

Build a coin-operated gumball machine. When you place a penny in the slot, the machine delivers a gumball to you. However, the gumball does not just plop out of the machine. Instead, it provides entertainment along the way such as playing sound effects, going down a series of chutes, turning a pinwheel—the wackier the better.

We will use marbles to represent the gumballs. For your machine, you may use an EV3, one or more sensors, and LEGO and non-LEGO building materials. If you need something, just ask!

This project is worth 100 points. The points will be awarded as shown in the chart below. Notice that you can earn as many as ten extra-credit points by successfully completing everything on the list.

Criterion	Possible points	Your points
Delivers gumball when activated	20	
Includes additional motion started by gumball	20	
Uses sensor	10	
Includes one simple machine	10	
Includes a second simple machine	10	
Activated by coin	10	
Makes music/sounds	10	
Works multiple times without being reset	10	
Delivers exactly one gumball whenever activated	10	

Teacher Information   Wacky Gumball Machine

The students design coin-activated gumball machines.

Objectives
1. To plan and carry out a major design project.
2. To engage in extensive testing and redesigning.

Materials
EV3, motors, sensors, LEGO® pieces, marbles or metal ball bearings, coins, computer, craft materials (optional)

Time: Several class periods

Notes
1. The nature of this project ensures that it will involve extensive testing and redesign. Allow plenty of time for testing and modifying the contraptions.
2. Designing a machine that will reliably deliver exactly one gumball every time is an especially challenging piece of this project. You may want to have the students share successful approaches with their classmates.
3. If they are available, you may wish to use metal ball bearings instead of marbles. Their heavier weight makes them more reliable triggers for mechanical devices within the machines.

Robo Artist

Build and program a robot artist that can turn a blank canvas into a beautiful piece of art. You may use the EV3, motors and sensors, LEGO pieces, tape, string, wire, paintbrushes, markers, crayons, etc. If you think of other materials you would like to use, just ask!

Your robot artist should use two different media (marker, pencil, crayon, paint, etc), two different techniques (drawing with a marker or brush, tapping, splattering, rolling, etc), two different simple machines, and a sensor. You should not get paint or marker anywhere but the paper—not on the floor, not on the walls, not on the EV3. Your picture should be large enough to cover much of the paper, rather than being squished into one corner.

Happy creating!

Critierion	Possible points	Your points
Uses one medium	10	
Uses a second medium	10	
Uses one technique	10	
Uses a second technique	10	
Uses a sensor	10	
Uses a simple machine	10	
Uses a second simple machine	10	
One change occurs during the program	10	
A second change occurs	10	
Picture confined to paper	10	
Picture in all four quadrants of paper	10	
No media on EV3, motors, sensors	10	
Total	120	

Teacher Information   Robo Artist

The students design robotic artists that draw and/or paint pictures.

Objectives
1. To plan and carry out a major design project.
2. To integrate art into a STEM activity, making it a STEAM activity.

Materials
EV3, motors, sensors, LEGO® pieces, large sheets of paper, paint, markers, other art supplies, computer

Time: Several class periods

Notes
1. The nature of this project ensures that it will involve testing and redesign. Allow time for testing and modifying the contraptions.
2. Supply plenty of practice paper so that the students can test and modify their drawing and painting mechanisms.
3. You may want to provide plastic wrap for covering the EV3 bricks, sensors, and motors.
4. You can have students draw random designs or challenge them to draw something specific, such as a person.

Part Seven: Low Tech Labs

These activities use simple LEGO® materials only. They do not include motors, sensors, or programming—just bricks. The activities explore a number of different topics in physics; the topics are listed in parentheses after each topic. The students gain a deeper understanding of the concepts by using them to solve problems. Along the way, the students are exposed to other important ideas, such as precision of measurement, experimental variables, and the engineering design process.

The activities in this section are:

1. Action/Reaction Car (distance, forces)
2. How Many Bricks in a Newton? (forces, mass, weight)
3. Gear Training (gears, rotation speed)
4. Worm Gears (gears, rotation speed)
5. Benham's Disks (gears, light, rotation speed)
6. Pulley Systems (mechanical advantage, pulleys)
7. Balancing Nails (center of gravity, equilibrium, stability)
8. Tightrope Walker (center of gravity, equilibrium, stability)
9. LEGO Balance (center of gravity, equilibrium, stability, torque)
10. Building Pressure (pressure)
11. Floating LEGO Bricks (density, mass, volume)
12. Cartesian Diver (density, pressure)

Action/Reaction Car

For every action, there is an equal and opposite reaction. For example, you can power a vehicle by attaching an inflated balloon to it. As air escapes out the back of the balloon, the vehicle is pushed forward.

Use this principle, Newton's third law of motion, to make a balloon-powered car that will travel as far as possible. Your car must be constructed of LEGO pieces and one balloon. You will make sketches of the initial and final versions of your car. After each trial, you will record how the car performed, one problem you encountered, and what modification you will make before the next trial.

Starter car:
Sketch:

Distance traveled:

Something that worked well:

Problem encountered:

Change to be made:

Second trial:
Distance traveled:

Something that worked well:

Problem encountered:

Change to be made:

Third trial:
Distance traveled:

Something that worked well:

Problem encountered:

Change to be made:

Continue to test and improve your car. At the end, make a sketch of your final car and note the farthest distance it traveled.

Final car:
Sketch:

Distance traveled:

Teacher Information   Action/Reaction Car

The students build balloon-powered cars.

Objectives
1. To design and build a balloon-powered car that demonstrates action and reaction forces.
2. To use the engineering design process.
3. To modify a design one step at a time.

Materials
LEGO® pieces, balloons, measuring tape

Time: Approximately 60 minutes

Notes
1. Like a rocket, the balloon-powered car demonstrates Newton's third law: For every action, there is an equal and opposite reaction.
2. You can either give the students one type of balloon or let them experiment with different shapes and sizes. Twelve-inch round balloons designed to hold helium work well.
3. Encourage the students to follow the instructions on the handout, making only one change at a time and then evaluating it. Many of the students will be tempted to implement several of their ideas at once. Point out to them that it will be much harder to determine the effect of each modification if they have made several changes at once. On the other hand, allow the students some leeway if they want to abandon their design and try a new one. Some of them will find that their first ideas do not work at all.
4. One of the biggest challenges in this activity is designing the frame that holds the end of the balloon so that it lets air out at a steady rate, not too fast and not too slow.
5. If the students are at a loss for changes to try, suggest that they experiment with different tires. Switching the tires can have a significant effect on the car's performance.

How Many Bricks in a Newton?

If you had a newton of two-stud-by-four-stud LEGO bricks, how many bricks would you have?

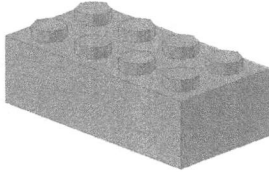

1. Make a guess (It does not matter whether you are correct.).

In this lab, you will figure out just how many LEGO bricks are in a newton.

First, find out how much the mass of a LEGO brick varies from one brick to another. Mass five LEGO bricks, one at a time. Record your results in the table below.

Brick number	Mass in grams
1	
2	
3	
4	
5	
Average	

2. Use the data that you collected to calculate how many LEGO bricks are in a newton. Write an explanation of your reasoning, showing the calculations that you performed and justifying each one.

Teacher Information   How Many Bricks in a Newton?

This activity gives the students a better sense of how large a newton is. It also gives them practice in using the relationship $F = ma$ to convert grams to newtons.

Objectives
1. To gain an understanding of the size of a newton.
2. To learn the difference between mass and weight and how to find one from the other.
3. To explain the reasoning behind a set of calculations.

Materials
2 x 4 LEGO® bricks, balance precise to at least hundredths of a gram

Time: Approximately 40 minutes

Notes
1. A 2 x 4 LEGO brick has a mass of around 2.2 grams, so there are roughly 45 bricks in a newton.
2. The students should be taught to use the balance before beginning this lab.
3. The students should be familiar with mass, weight, and Newton's second law before doing this lab.
4. You may want to discuss two different methods of finding the average mass of the bricks—massing them one at a time as in this activity or massing all five together. Massing the five together will reduce the error in massing (since you are massing once instead of five times), but it does not give you an idea of how much the mass varies from brick to brick.
5. If the students are familiar with the concept of significant figures, this activity is a good place to discuss them.

Answers to How Many Bricks in a Newton?

1. Answers will vary.

   Sample data:

Brick number	Mass in grams
1	2.22
2	2.22
3	2.19
4	2.22
5	2.19
Average	2.21

2. The average mass of the five bricks was 2.21 grams. In order to convert the mass to kilograms, divide by 1000. So, the mass in kilograms is 0.00221. To find the weight of one brick in newtons, multiply the mass in kilograms by the acceleration due to gravity, 9.8 m/s/s.

   (0.00221 kg) (9.8 m/s/s) = 0.022 newtons.

   Since each brick weighs an average of 0.0022 newtons, divide one newton by this average weight to determine how many bricks are in a newton:

   1/ (0.022) = 45.

   There are approximately 45 LEGO bricks in one newton.

Gear Training: Exploring the Basics of Gears

First, some terms: Two or more gears meshed together are called a **gear train**. The gear to which the force is initially applied is called the **driver**. The final gear is called the **follower**, or **driven gear**. Any gears between the driver and the follower are called **idlers**.

Now, try building each of the trains described below. Sketch or list the gears you used for each. You may use three types of LEGO spur gears for this challenge: the small gear with eight teeth, the medium gear with twenty-four teeth, and the large gear with forty teeth.

1. The driver and the follower turn at the same speed, but in opposite directions.

2. The driver and the follower turn at the same speed in the same direction.

3. The driver turns three times as fast as the follower.

4. The follower turns five times as fast as the driver.

5. The follower turns twenty-five times as fast as the driver. (There's a trick to this one!)

Gearing up and gearing down: **Gearing up** means that the follower in a particular gear train turns faster than the driver. **Gearing down** means that the follower turns more slowly than the driver.

6. Look back at the gear trains you made. For each one, note whether it was gearing up, gearing down, or neither.

Teacher Information   Gear Training

The students explore gears, building gear trains and learning about ratios, gearing up, and gearing down.

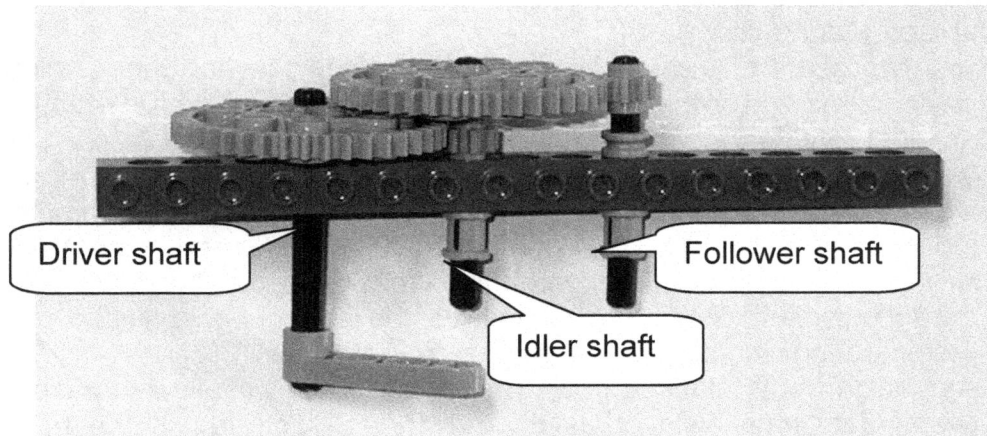

The eight-tooth follower in this gear train turns twenty-five times faster than the driver and its crank.

Objectives
1. To build a variety of gear trains.
2. To calculate gear ratios.
3. To understand gearing up and gearing down.

Materials
LEGO® parts including gears

Time: Approximately 30 minutes

Notes
1. Provide long beams, axles, bushings, and a variety of gears for each group.
2. The eight-, twenty-four-, and forty-tooth gears work well for this activity because they can be easily meshed along a single beam.
3. The EV3 set contains two forty-tooth gears, four twenty-four-tooth gears, and four eight-tooth gears—enough equipment to complete this challenge without using additional materials.

Answers to Gear Training:
Exploring the Basics of Gears

1. Any two gears of the same size, meshed together, will turn at the same speed but in opposite directions. For example, two forty-tooth gears. (Neither gearing up nor gearing down)
2. Any two gears of the same size, with an idler between them, will turn at the same speed and in the same direction. For example, two forty-tooth gears with an eight-tooth gear between them. (Neither gearing up nor gearing down)
3. Any gear train where the follower is three times as large as the driver will make the driver turn three times as fast as the follower. For example, an eight-tooth gear meshed with a twenty-four-tooth gear. (Gearing down)
4. Any gear train where the driver is five times as large as the follower will make the follower turn five times as fast as the driver. For example, a forty-tooth gear meshed with an eight-tooth gear. (Gearing up)
5. A gear train with stacked gears on an idler shaft will allow the follower to turn twenty-five times as fast as the driver, as illustrated in the picture at the beginning of the Teacher Information page. The first axle holds a forty-tooth gear. The second axle contains two gears, an eight-tooth gear meshed with the large gear on the first axle and a forty-tooth gear. The forty-tooth gear on the second axle is meshed with an eight-tooth gear on the third axle. Each pair of gears increase the speed by five times, so the two pairs together give an increase of twenty-five times. (Gearing up)
6. The answers are given in parentheses after each of the above questions.

Worm Gears

A worm gear has a different shape than other gears you have seen. It is shaped like a corkscrew. Let's investigate how it works.

Build a frame to mesh a worm gear with an eight-tooth spur gear. Attach a crank to each gear.

1. Try turning the crank attached to the worm gear. How many times must you turn it in order for the eight-tooth gear to make one complete rotation?

2. Examine a worm gear carefully. How many teeth does it have?

3. Now try turning the crank attached to the eight-tooth gear. How many times must you turn it to make the worm gear turn once?

4. You wish to construct a gear train with a slowdown of 40 times—the driver must turn 40 times in order for the follower to turn once. Describe one possible solution.

Extra credit: Describe a different gear train that would also give you a slowdown of 40 times.

Teacher Information   Worm Gears

This activity introduces the students to worm gears.

Objectives
1. To learn how to use a worm gear.

Materials
Worm and spur gears, LEGO® pieces to make a frame

Time: Approximately 30 minutes

Notes
1. You may want to build an example frame for the students to use as a model. One possibility is shown below.

2. The handout includes a trick question—the third one. The students are asked to turn the crank on the spur gear in order to turn the worm gear. Of course, since the worm gear is a one-way gear, the gears lock rather than turning. Trying it for themselves helps the concept to register more than if the students are simply told that a worm gear cannot be used for gearing up.

Answers to Worm Gears

1. The worm gear must turn eight times in order for the 8-tooth gear to make one complete rotation.
2. A worm gear has only one tooth, which spirals the length of the gear.
3. This is a trick question. The 8-tooth gear cannot be turned. The worm gear can only be used for gearing down.
4. Two possible solutions (the second solution can be used for the extra-credit question):
    a. Mesh a worm gear with a forty-tooth gear.
    b. Mesh a worm gear with an eight-tooth gear, and then attach another gear train on the same axle, consisting of an eight-tooth gear meshed with a forty-tooth gear.

Benham's Disks

Build a gear train where the follower rotates 25 times faster than the driver. Cut out one of the disks from the design sheet and glue it to thin cardboard. Mount the disk above the follower, anchoring it to the axle with bushings.

1. Now spin the driver at varying speeds and watch the disk. What colors do you see?

A number of black-and-white patterns fool your eye into seeing flashes of color when they are spun. Try the other design from the sheet, and then invent your own pattern.

2. Sketch any of your invented patterns that produce colors.

3. Make a gear train in which the follower spins 125 times faster than the driver. Use it to spin your disk. What happens?

4. Divide a circle into eight equal wedges. Leave one white and color the other seven the colors of the rainbow: red, orange, yellow, green, blue, indigo, violet. Spin the circle. What do you see?

Benham's Disks Design Sheet

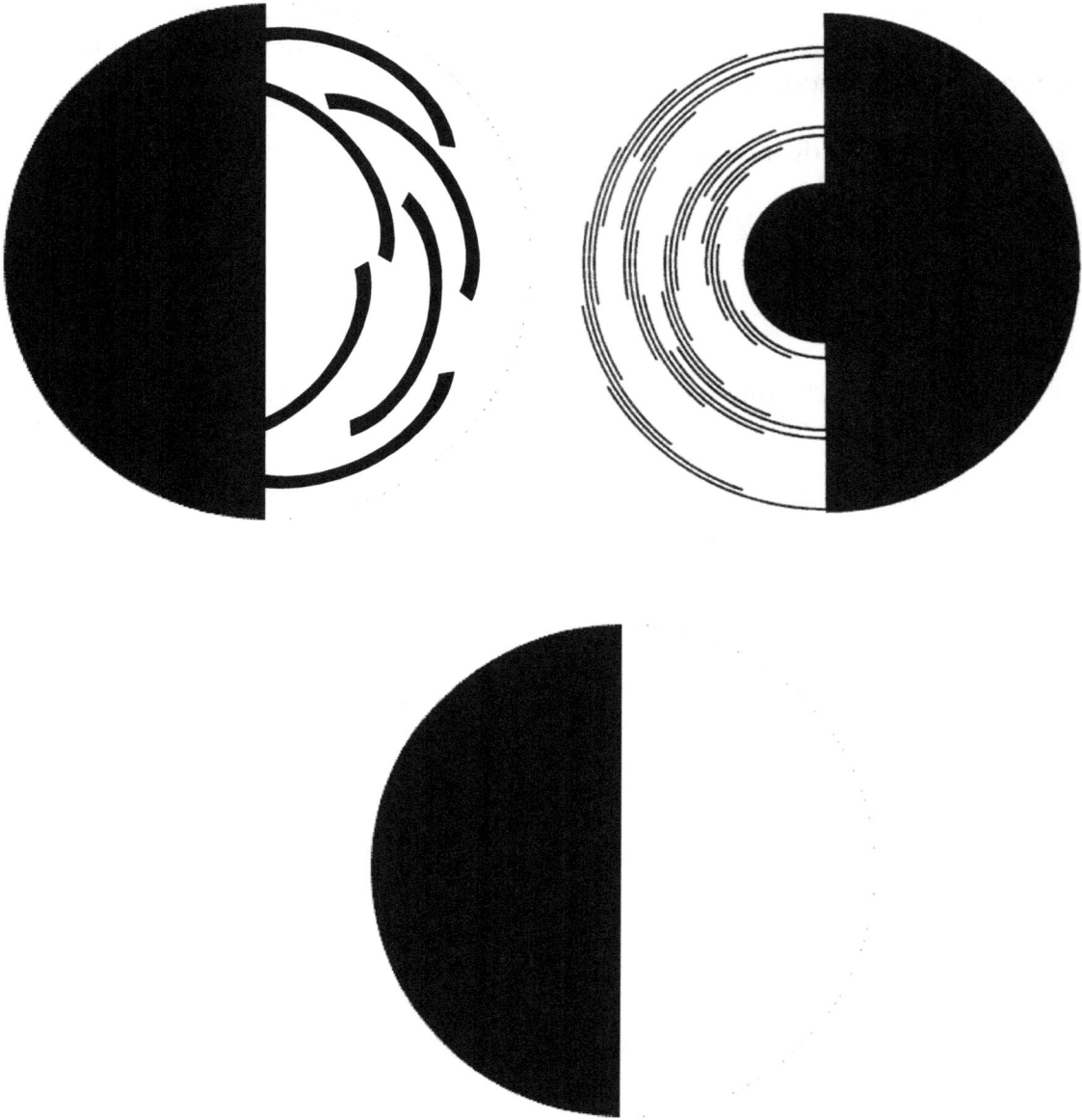

Teacher Information   Benham's Disks

The students use gears to spin Benham's disks.

Objectives
1. To build a gear train in which the follower moves 25 times faster than the driver.
2. To experiment with Benham's disks and the color spectrum.

Materials
LEGO® pieces, thin cardboard, scissors, glue, markers, compasses for making circles

Time: Approximately 30 minutes

Notes
1. This activity makes a good follow-up activity to Gear Training. The students who finish Gear Training quickly will have more time to design their own Benham's disks.
2. The third template, the half-black/half-white disk, can be used for the students' own designs.
3. Benham was a toymaker in the 1800's who discovered this illusion while experimenting with tops. The effect is produced by differences in the cones of your eyes. You have three types of cones, one for detecting blue light, one for red, and one for green. The three types of cones have different response patterns, so that the flashes of white light on the spinning disk activate them in different ways. Their varying responses create the illusion of color.

Answers to Benham's Disks

1. Answers will vary. Often, the colors seen will vary with the speed of rotation. Have the students experiment with different speeds.
2. Answers will vary. Generally, patterns that are at least half black and contain thin black arcs are the most successful.
3. This gear train will be difficult, if not impossible, to turn because increasing the speed of the gears reduces the torque.
4. The spinning rainbow will produce a white color (or something close to it).

Pulley Systems

Different pulley systems have different ideal mechanical advantages, depending upon their configurations. Can you construct a pulley system (using as many or as few pulleys as you wish) with an ideal mechanical advantage of one? Of two? Of six?

Construct a pulley system for each of the following ideal mechanical advantages. Once your pulley system is complete, make a sketch of it, then have the teacher check it to make sure it is correct.

I.M.A.	Sketch of pulley system	Teacher check
1		
2		
3		

4		
5		
6		

Teacher Information   Pulley Systems

This activity allows the students to explore pulley systems and to calculate the mechanical advantage for different systems.

Objectives
1. To rig pulley systems.
2. To find the ideal mechanical advantage of various systems.

Materials
LEGO$^®$ pieces including pulleys, string, weighted brick or mass to be lifted

Time: Approximately 40 minutes

Notes
1. The students should have some understanding of mechanical advantage before beginning the activity.
2. Encourage the students to begin by building a frame to support the pulleys. Doing so will making rigging the more complicated systems easier.
3. If you wish, you can award points or prizes at the end based upon the number of pulley systems completed correctly.

Answers to Pulley Systems

1. Single fixed pulley
   Ideal mechanical advantage of one

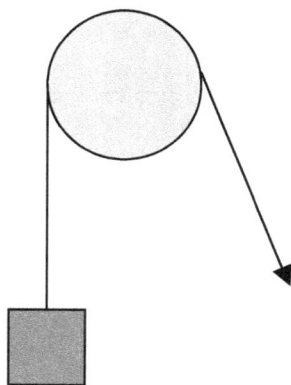

2. One fixed pulley, one movable pulley
   Ideal mechanical advantage of two

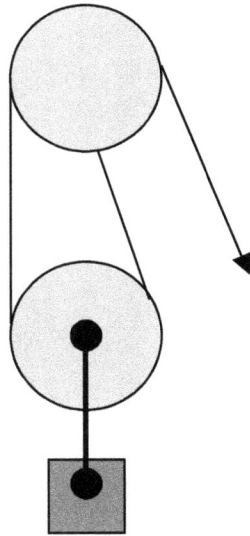

3. One fixed pulley, one movable pulley
   Ideal mechanical advantage of three

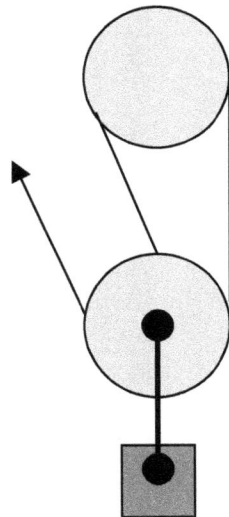

4. Two fixed pulleys, two movable pulleys
   Ideal mechanical advantage of four

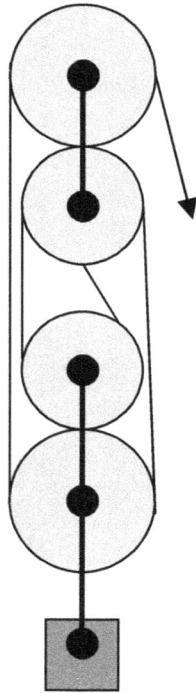

5. Two fixed pulleys, two movable pulleys
   Ideal mechanical advantage of five

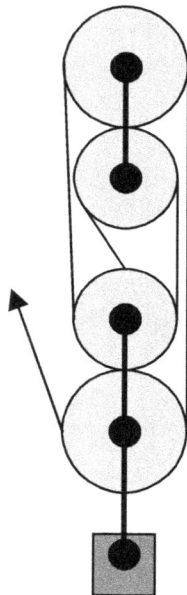

6. Three fixed pulleys, three movable pulleys
   Ideal mechanical advantage of six

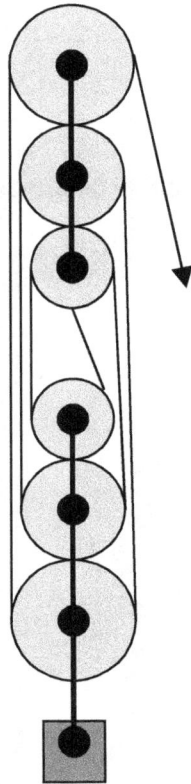

Balancing Nails

It is quite easy to balance a single LEGO nail on a stand. Can you find a way to balance more?

Part I: Balancing Act

Balance as many nails as you can.

1. How many nails were you able to balance?

2. Make a sketch of your arrangement below.

Part II: Long and Short

Once you know the trick for balancing eight nails, try balancing two different lengths of nails. First use twelve-stud-axle nails. To balance the nails, first arrange them on the ground, then lift them as a group and place them on the stand.

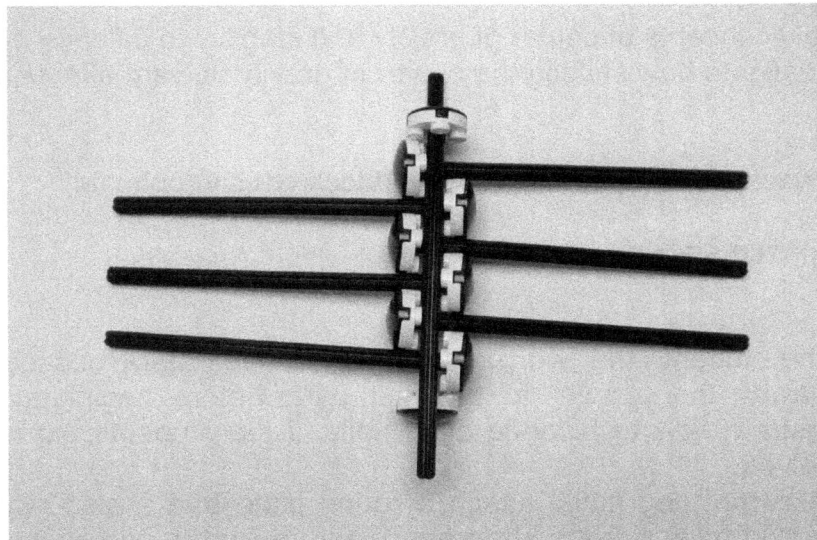

Once you are successful at balancing the long nails, replace the six side nails with six-stud-axle nails and try again.

3. Was it harder to balance the long nails or the short nails?

4. Explain your results in terms of center of gravity.

Teacher Information   Balancing Nails

The students balance LEGO® nails.

Objectives
1. To apply concepts of center of gravity and stability to balance nails.
2. To investigate how shifting the center of gravity upward affects stability.

Materials
LEGO® pieces, nails (optional), wooden block or clay (optional)

Time: Approximately 40 minutes

Notes
1. Give the students the two parts of this lab separately, because the second part gives away the answer to the first part. Before beginning the second part, demonstrate how to balance eight nails, if the students did not figure it out for themselves.
2. To make the long nails, attach a round plate and a skid plate to the top of a twelve-stud axle. For the short nails used in Part II, use six-stud axles. To make the stand, mount a twelve-stud axle vertically. Place a bushing on the top. (See the photograph below.)

The stand, a short nail, and a long nail.

3. To balance the nails, first arrange them on the ground, then lift them as a group and place them on the stand. (See the photograph below.)

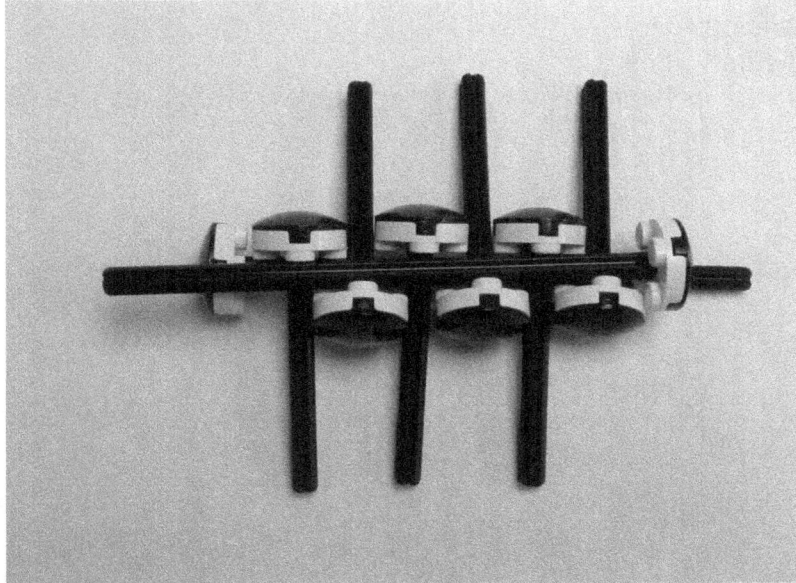

Short nails arranged on the ground, ready to be lifted onto the stand.

4. Before starting this project, make sure that the students have an understanding of center of gravity and stability. In particular, the students should understand that an object is at stable equilibrium if its center of gravity is below its balance point. They should also realize that the center of gravity of an object does not have to part of the object.
5. Part I of this activity can also be done with real nails.

Answers to Balancing Nails

Part I: Balancing Act
1. Answers will vary.
2. Answers will vary. The photograph below shows an arrangement for balancing eight nails.

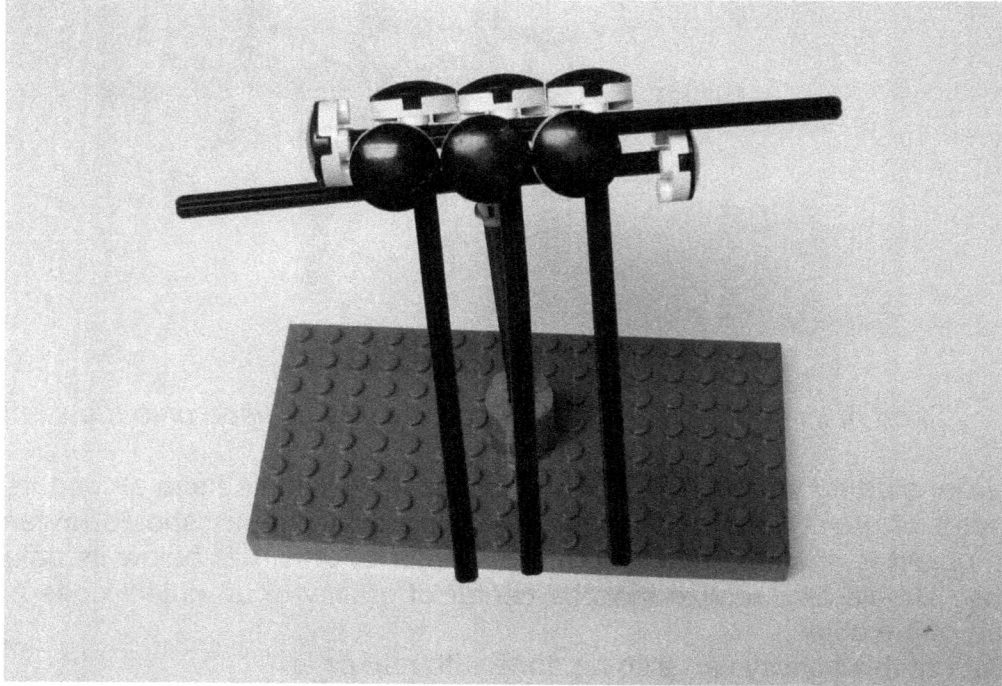

Part II: Long and Short
3. The longer nails are easier to balance.
4. The arrangement of the longer nails has a lower center of gravity. Because its center of gravity is below the balancing point, this arrangement is stable.

Acknowledgement: This activity is based upon an activity at the Museum of Science, Boston.

Tightrope Walker

Walking on a tightrope isn't always as difficult as it looks. Often, the performers find ways to lower their center of gravity in order to make balancing easier. In fact, if the walker's center of gravity is below the rope, he or she will tend to return to an upright position if disturbed.

Build your own LEGO tightrope walker, at least ten centimeters tall. The walker will be mounted on a pulley. The pulley will roll along a string stretched taut and anchored at the ends. Both the pulley and the walker must remain above the rope. However, the walker can have arms, legs, or accessories that dangle below the rope. He or she can even be holding onto a person or animal that dangles below the rope.

1. Make a sketch of your walker below.

2. Place an X at its approximate center of gravity.

Teacher Information   Tightrope Walker

The students build tightrope walkers out of LEGO® pieces.

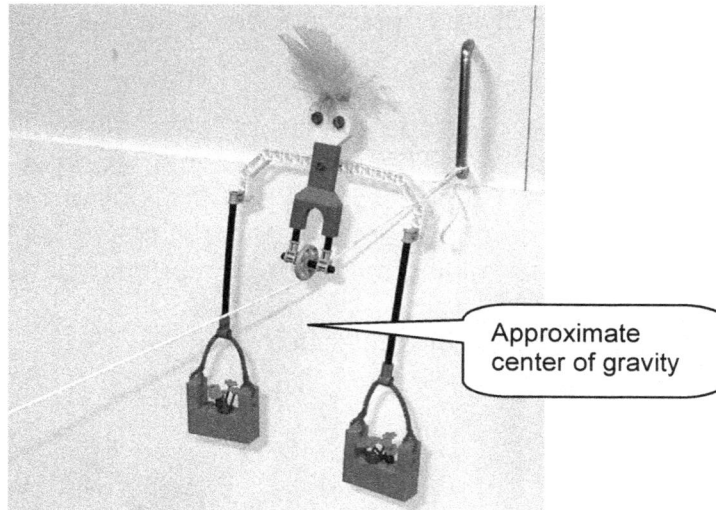

Approximate center of gravity

Objectives
1. To apply concepts of center of gravity and stability to build a LEGO tightrope walker.
2. To use the engineering design process.

Materials
LEGO pieces, string, craft materials

Time: Approximately 40 minutes

Notes
1. Before starting this project, make sure that the students have an understanding of center of gravity and stability; in particular, that an object is at stable equilibrium if its center of gravity is below its balance point. They should also realize that the center of gravity of an object does not have to be on a physical part of the actual object. For example, the center of gravity of a hoop is at the center of the hoop.
2. Getting the tightrope walkers to balance will probably require several rounds of adjustment after the completed figures are placed on the tightrope. Encourage the students to make small adjustments to their figures rather than major changes.

Answers to Tightrope Walker

1. Answers will vary.
2. If the object is stable, the center of gravity should lie directly underneath the figure, below the tightrope.

LEGO Balance

Build your own pan balance out of LEGO pieces, string, and a set of standard metric masses.

Here are your design specifications:

•Your balance must be able to mass objects between 0 and 50 grams.

•Your balance must be able to measure to within 0.5 grams of the actual mass of the object. (You're welcome to make it more precise, of course.)

Other than these two constraints, the design is up to you. After you have built your balance, you may test it (and make improvements to it) as many times as you wish. A regular balance is available so that you can compare the results from it with those from your LEGO balance.

When you are confident that your balance can measure to within 0.5 grams of the actual mass, you are ready for the official test. You will be given a basket containing four objects, labeled 1, 2, 3, and 4, and asked to find the mass of each one.

Good luck!

Balance Test

You will be given a basket containing four objects, labeled 1, 2, 3, and 4. Mass each object and write down the answer below.

Object Number	Mass in grams
1	
2	
3	
4	

Teacher Information   LEGO Balance

The students build LEGO® balances that are accurate to within 0.5 grams.

The weighted bricks on the arms of this balance help it to be self-leveling by lowering the center of gravity below the place where the arms are attached to the base.

Objectives
1. To apply concepts of center of gravity and torque to build a LEGO balance.
2. To grapple with precision of measurement and interpolation.
3. To use the engineering design process.

Materials
LEGO pieces, set of metric masses, string, pan balance, electronic or triple-beam balance, objects for massing

Time: Approximately 120 minutes

Notes
1. At first, this project can seem daunting to students, since they are being asked to construct a precise and accurate balance out of LEGO pieces. However, the nature of the task also gives them a real sense of accomplishment when they complete it successfully. The way in which the activity is structured virtually guarantees their success.
2. Make sure that the students are familiar with pan balances before starting this project. Let them use a pan balance to mass a few objects to get a solid sense of how the balance works. It is also helpful to have different types of pan balances,

or pictures of them, so that students may see some of the various ways of attaching the pans to the arm.

3. Generally, the smallest mass in a metric set is one gram, so students will have to use interpolation to mass objects to within one-half gram.

4. Designing a balance requires an understanding of torque (or levers), since an object further out along an arm of the balance will exert a greater torque than one close to the pivot point. The students can make use of this concept to increase the precision of the balance by making the arms longer.

5. A common design flaw in the students' balances is to have the center of gravity above the balance point. In this case, the arms tend to swing wildly if disturbed, rather than coming to rest in a horizontal position. Adding a pair of weighted bricks to the underside of the arms will lower the center of gravity and improve the performance of the balance. If a group of students is experiencing the problem of an unstable balance, try turning their balance upside down. Then, the center of gravity is below the balance point, the moving part of the balance is at stable equilibrium, and the balance behaves much better. You can leave the students to figure out why turning the balance upside down helped. Ask them questions to get them thinking along the right lines if necessary.

6. Another common design flaw is to anchor the pans to the arms at more than one point. If the students do this, their balance will behave like a seesaw, with objects placed at the outer edge of the pan appearing to have more mass than those placed on the inner edge.

7. In general, the easiest design to make work is one with high arms and pans suspended by strings.

8. The assessment for this activity shifts much of the responsibility to the students, since they are deciding when their balance is ready to be tested. Before giving them the bag of unknown objects to test, make sure that they have tested their balance on a number of practice objects, covering the full range of masses between zero grams and fifty grams.

9. When the students bring their completed test to you, you may want to scan it before grading it to make sure that their answers are in the right ballpark. On a few occasions, I have had a group submit a test with inaccurate answers. I hand the test back to them without grading it and look at their balance with them to try to pinpoint the problems. After they have modified the balance and conducted more trials, I give them a different basket of unknowns for their test.

Building Pressure

Make a stack of eight two-stud-by-four-stud LEGO bricks. Set the stack on a table. The stack of bricks is exerting pressure on the table—the weight of the eight bricks pushing down on the area of the table covered by one brick, or a ratio of 8:1.

1. Using all eight LEGO bricks, rearrange them so that they are exerting only half as much pressure on the table as the original stack (in other words, you want your new grouping to have a ratio of 4:1). Sketch your arrangement.

2. Split your eight bricks into two groups of four. Arrange each group so that it exerts a different amount of pressure on the table than the other group of four does. Sketch your arrangement.

3. Next, split the eight bricks into two groups of unequal size, but arrange each group so that it exerts the same pressure on the table. Sketch your arrangement.

4. Arrange the eight bricks so that they exert as low a pressure on the table as possible. Sketch your arrangement.

5. Arrange the eight bricks so that they exert as high a pressure on the table as possible. (The bottom of the stack must rest flat against the table.) Sketch your arrangement.

Teacher Information   Building Pressure

This activity gives the students a chance to experiment with the concept of pressure.

Objectives
1. To gain an understanding of the concept of pressure.
2. To find creative solutions to problems.

Materials
2 x 4 LEGO® bricks

Time: Approximately 30 minutes

Notes
1. Give the students LEGO bricks to manipulate as they work on the activity. Each student or group of students will need eight bricks.
2. If you wish, you may have the students mass the bricks and measure their dimensions, so that they will be able to calculate the pressure in standard units, rather than using ratios.

Answers to Building Pressure

1. One possibility is to place two stacks of four bricks each side by side.

2. There are many possible answers. One is to form one group of four bricks into a single stack and the second group into two stacks of two. The tall stack has a brick/area ratio of 4:1. The other group has a ratio of 2:1.

3. Again, there are many possibilities. For example one group could be made up of a single stack of bricks two high, while the other group could contain three stacks of two bricks. Both groups have a ratio of 2:1.

4. Spread the bricks out in a single layer, with a ratio of 1:1.

5. Place one brick on its end and attach the other seven to it. Since the end of the brick has about one-third the area of the bottom of the brick, this arrangement has a brick/area ratio of roughly 24:1.

Allowing the bricks to hang over the edge of the table leads to even better solutions (and a chance to discuss center of gravity).

Floating LEGO Bricks

How dense is a LEGO weighted brick? To find out, you will first need to find the mass and volume of the brick.

1. Use a balance to find the mass of the weighted brick. Record your answer in grams.

2. Find the volume of the brick, either by displacement or by measuring the length, width, and height and multiplying them together. Record your answer in cubic centimeters.

3. Find the density of the brick by dividing the mass by the volume. Record your answer.

Now, your challenge is to make your weighted brick float by adding pairs of 2 x 8 unweighted bricks to the top of it.

4. These pairs of bricks are the same volume as the weighted brick, though not the same mass. Use the balance to find the mass of a pair of bricks. Next, find the density of the pair. Record your answers.

5. Before you try adding any pairs of bricks to your weighted brick, calculate how many pairs you will need. (Remember, the density of water is 1.0 g/cm^3.) Show your calculations below.

6. Once you have calculated how many bricks you will need, build your stack and try floating it. Did it float?

7. Answer 7a if your stack floated and 7b if it did not.
   a. If your stack floated, remove a pair of bricks and try again. If the new stack still floats, calculate its density. Show your work.

   b. If your stack did not float, add pairs of bricks until the stack floats. Find the density of the new stack. Show your work.

Teacher Information   Floating LEGO Bricks

The students find the density of a weighted brick, and then calculate how many unweighted bricks must be placed on top of it in order for the entire stack to float.

Objectives
1. To use a balance to find mass.
2. To find volume, either by displacement or by measuring the dimensions of the object.
3. To find the density of an object.
4. To understand how density and buoyancy are related.
5. To use density to solve a problem.

Materials
Electronic balance precise to at least hundredths of a gram, weighted brick, metric ruler, 2 x 8 LEGO® bricks, container at least 15 cm tall for floating stack, container for measuring volume by displacement (optional)

Time: Approximately 40 minutes

Notes
1. The students should be familiar with the concepts of mass, volume, and density before beginning this activity. They should also understand how to use an electronic balance.
2. In creating the stack, the weighted brick must be at the bottom of the stack, so that air is trapped inside the unweighted bricks.
3. The floating stack is approximately 14 cm high.

Answers to Floating LEGO Bricks

1. The mass of a weighted brick is generally around 53 grams.
2. The dimensions of the weighted brick are approximately 4.8 cm by 1.5 cm by 1.9 cm. Its volume is around 14 cm^3.
3. Assuming a mass of 53 grams and a volume of 14 cm^3, the density of the weighted brick is around 3.8 g/cm^3.
4. The mass of a pair of 2 x 8 bricks is around 7.0 grams. The volume is the same as that of the weighted brick, around 14 cm^3. The density of the pair is approximately 0.5 g/cm^3.
5. To make the stack float, its density must be less than 1.0 g/cm^3. If you add six pairs of 2 x 8 bricks to the weighted brick, the total mass of the stack will be approximately 42 + 53 = 95 grams. The total volume of the stack will be approximately 7 x 14 = 98 grams. With a density of 0.97 g/cm^3, this stack should float. Depending upon the mass of the weighted brick used, the number of 2 x 8 bricks needed may vary slightly.

6-7. Answers will vary.

Cartesian Diver

A Cartesian diver is an object that floats in a sealed bottle until the bottle is squeezed, at which point the object sinks. When the pressure on the bottle is released, the object rises back to the top. Make a Cartesian diver out of an eyedropper.

1. What changes do you see inside the dropper as it goes up and down?

Now you are ready to design your own diver. Make a diver that is able to go up and down, using the materials provided.

2. Sketch your successful diver below, labeling the materials used.

Use what you have learned to make a LEGO diver—a minifigure that goes up and down. You may use other LEGO pieces and/or the other materials provided.

3. Sketch your diver, labeling the materials used.

4. Based upon your experimentation, what properties must an object have in order to be a Cartesian diver?

Teacher Information   Cartesian Diver

This activity gives the students a chance to experiment with the concepts of buoyancy, density, and pressure.

Objectives
1. To gain an understanding of the concepts of buoyancy, density, and pressure.
2. To find creative solutions to problems.

Materials
One-liter plastic soda bottle with lid. glass eyedropper with rubber bulb, LEGO® minifigures, materials for making Cartesian divers (see note #1 below)

Time: Approximately 60 minutes

Notes
1. A variety of compressible materials can be used to create Cartesian divers. Cranberries, pieces of mushroom, and small marshmallows (though they disintegrate over time) are foods that work well. Non-food possibilities include Styrofoam, cork, and uninflated balloons. In addition, you will need heavier materials to provide weight, such as paper clips, pipe cleaners, and metal washers.
2. To turn the eyedropper into a Cartesian diver, fill it partly full of water so that it barely floats. As they squeeze the bottle, the students will be able to see the water level rise in the dropper.
3. To turn the other materials into Cartesian divers, combine a compressible material, such as a small marshmallow, with enough added weight so that the combination barely floats.
4. To turn a minifigure into a Cartesian diver, it may be necessary to add a floating compressible material and/or a weight. Some minifigures, such as the ghost, will trap enough air that they do not need any added compressible material. Some minifigures will be dense enough that they need little or no added weight.
5. Depending upon the size of the bottle opening, you may need to raise the minifigure's arms over its head in order to fit the figure into the bottle.

Answers to Cartesian Diver

1. The water level rises inside the dropper as it goes to the bottom of the bottle and drops as the dropper returns to the surface.

2-3. Answers will vary.

4. To be a Cartesian diver, an object needs to include a compressible material. It also needs to barely float.

Appendix A: Activities Listed by Topic

For ubiquitous programming blocks, such as Wait and Loop, only introductory activities or ones that introduce new ideas are listed.

Acceleration
    Crossing the Lines 5-9
    Deriving 5-16

Algorithm
    Outside the Box 4-18
    Tabletop Treasure 4-21
    Loaded Dice 4-58
    Grassfire 4-76

Angle
    Pandora's Box 2-8
    Spirographer 4-24
    Ramp Up 4-42
    Peak Performance 4-46

Area
    Cloverleaf 4-16

Array
    Brick-button Navigator 3-41

Bayes' rule
    Which Room? 5-28

Biology
    Reaction Time 4-73
    Robotic Zoo 6-19
    Benham's Disks 7-14

Brick buttons
    Red, Red, Red 3-4
    Roll of the Die 3-31
    Dog Year Converter 3-34
    Touch Tally 3-39
    Brick-button Navigator 3-41
    Random or Not 4-54
    Loaded Dice 4-58

Brick Status Light block
    Traffic Light 3-24
    Intersection 3-27

Buoyancy
    Floating LEGO Bricks 7-38
    Cartesian Diver 7-41

Center of gravity
    Ramp Up 4-42
    Peak Performance 4-46
    Balancing Nails 7-23
    Tightrope Walker 7-28
    LEGO Balance 7-30

Circumference
    Driving 5-12

Color sensor
    See and Say 3-3
    Red, Red, Red 3-4
    Red or Not 3-5
    Rainbow Detector 3-6
    Cockroach 3-12
    Daytime Fan 3-18
    Outside the Box 4-18
    Bug Battle 4-18
    Tabletop Treasure 4-21
    Musical Instrument 4-51
    Light and Dark Scavenger Hunt 5-2
    Bragging Rights 5-4
    Thunderstorm 5-6
    Crossing the Lines 5-9
    Which Room? 5-28

Connector pegs
    Making Connections 2-2
    Build a Box 2-4

Appendix B: Mindstorms Equipment Used for Each Activity

	EV3	Large motor	Medium motor	Touch sensor	Color sensor	Gyro sensor	Rotation sensor	Ultrasonic sensor	Sound Sensor	Temperature Sensor
INTRODUCTORY ACTIVITIES										
Making Connections										
Build a Box	X									
Fancy Box	X									
Pandora's Box	X					X				
Sneak Attack	X					X				
Baker's Dozen Car	X	X								
PROGRAMMING SEQUENCES										
See and Say	X				X					
Red, Red, Red	X				X					
Red or Not	X				X					
Rainbow Detector	X				X					
Lurch	X	X					X			
Snake	X	X								
Cockroach	X	X			X					
Unsynchronized Motors	X	X		X			X			
Push-button Fan	X		X	X						
Daytime Fan	X		X		X					
Gradual Fan	X		X							
Five-speed Fan			X	X						
Traffic Light	X									
Walk Signal	X				X					
Animated Walker	X				X					
Intersection	X				X					
Roll of the Die	X									

X indicates possible use.

	EV3	Large motor	Medium motor	Touch sensor	Color sensor	Gyro sensor	Rotation sensor	Ultrasonic sensor	Sound sensor	Temperature sensor
Even or Odd	X									
Countdown	X									
Dog Year Converter	X									
Cautious Car	X	X		X						
Touch Tally	X			X						
Brick-button Navigator	X	X								
Mail Delivery	X	X								
STEM ACTIVITIES										
Getting Up to Speed	X	X								
Stop for Pedestrians	X	X								
Parking Space	X	X								
No Wheels	X	X								
At a Snail's Pace	X	X								
Cloverleaf	X	X								
Outside the Box	X	X			X					
Bug Battle	X	X			X					
Tabletop Treasure	X	X			X			X		
Spirographer	X	X				X				
Puppy Bot	X	X	X					X		
Range Puppy	X	X	X					X		
Proportional Puppy	X	X	X					X		
Haunted House	X	X			X	X	X			
Clean Sweep	X	X		X						
Perfect Pitcher	X	X		X						
Different Drummer	X	X	X							
Ramp Up	X	X				X				

X indicates possible use.

	EV3	Large motor	Medium motor	Touch sensor	Color sensor	Gyro sensor	Rotation sensor	Ultrasonic sensor	Sound sensor	Temperature sensor
Peak Performance	X	X				X				
Applause Meter	X		X						X	
Musical Instrument	X				X			X		
Random or Not	X									
Loaded Dice	X									
Efron's Dice	X			X						
Voting Machine	X			X						
Do You Have a Sister?	X			X						
Reaction Time	X			X						
Grassfire	X	X				X	X			
DATA LOGGING ACTIVITIES										
Light and Dark Scavenger Hunt	X				X					
Bragging Rights	X				X					
Thunderstorm	X				X				X	
Crossing the Lines	X	X			X					
Driving	X	X					X			
Deriving	X	X					X			
Zigzag and Diamond	X	X					X			
Puppy Data Logging	X	X	X					X		
Which Room?	X	X			X					
Stir It Up	X	X								X
It's a Breeze	X									
Cool It Fast	X	X								X
ENGINEERING PROJECTS										
Meet and Greet	X	X	X	X	X	X		X		
Fairytale Fix: Goldilocks	X				X	X	X			

X indicates possible use.

	EV3	Large motor	Medium motor	Touch sensor	Color sensor	Gyro sensor	Rotation sensor	Ultrasonic sensor	Sound sensor	Temperature sensor
Fairytale Fix: Rapunzel	X	X	X	X	X	X	X	X	X	
Fairytale Fix: Cinderella	X	X	X	X	X	X	X	X	X	
Dancing Bot	X	X	X				X			
Household Helper	X	X	X	X	X	X	X	X	X	X
Music Box	X	X	X				X			
Mini Golf	X	X	X	X	X	X	X	X	X	
Robotic Zoo	X	X	X	X	X	X	X	X	X	
Chain Reaction Machine	X	X	X	X	X	X	X	X	X	
EGGcellent Contraption	X	X	X	X	X	X	X	X	X	
Wacky Gumball Machine	X	X	X	X	X	X	X	X	X	
Robo Artist	X	X	X	X	X	X	X	X	X	
LOW TECH LABS										
Action/Reaction Car										
How Many Bricks in a Newton?										
Gear Training										
Worm Gears										
Benham's Disks										
Pulley System										
Balancing Nails										
Tightrope Walker										
LEGO Balance										
Building Pressure										
Floating LEGO Bricks										
Cartesian Diver										

X *indicates possible use.*

www.ingramcontent.com/pod-product-compliance
Lightning Source LLC
Chambersburg PA
CBHW081806200326
41597CB00023B/4168